Chemical Engineers in Medicine 2019

Topical Conference at the 2019 AIChE Meeting

Orlando, Florida, USA
10-15 November 2019

ISBN: 978-1-7138-0563-2

Printed from e-media with permission by:

Curran Associates, Inc.
57 Morehouse Lane
Red Hook, NY 12571

Some format issues inherent in the e-media version may also appear in this print version.

Printed with permission by Curran Associates, Inc. (2020)

For permission requests, please contact AIChE
at the address below.

AIChE
120 Wall Street, FL 23
New York, NY 10005-4020

Phone: (800) 242-4363
Fax: (203) 775-5177

www.aiche.org

Additional copies of this publication are available from:

Curran Associates, Inc.
57 Morehouse Lane
Red Hook, NY 12571 USA
Phone: 845-758-0400
Fax: 845-758-2633
Email: curran@proceedings.com
Web: www.proceedings.com

TABLE OF CONTENTS

555720 Self-Assembling Nanotechnology for Cancer Theranostics: From Computer-Assisted Design to *In Vivo* Applications

As a part of a pan-European task force in the field, our team has developed a series of innovative theranostic systems which proven to be excellent agents in cancer imaging and therapeutics in vivo. Specifically, exploiting the quintessence of nanotechnology, i.e., the self-assembling process of small, amphiphilic molecules, in the last decade we created a series of high-performance, non-toxic nanosized micelles that, depending on the specific chemistry, are able to (e.g.):

• encapsulate anticancer drugs with high loading capacity, enhance drug potency and combat drug resistance by promoting cellular uptake whilst decreasing drug efflux;

• perform specific and effective gene silencing via targeted small interfering RNA (siRNA) delivery;

• provide PET images with significantly superior imaging quality relating to sensitivity, specificity and accuracy when compared to the clinical standard [^{18}F]FDG.

In this contribution we will report in details the full pathways leading to these three types of nanosystems, from their computer assisted molecular design to their in vivo performance.

References

Pricl et alì., Proc. Natl. Acad. Sci. USA, 115 (2018) 11454-11459; J. Am. Chem. Soc., 140 (2018) 16264-16274; Dagrada et al., Mol. Pharm. 15 (2018) 4689-4701; Chen et al., Small 12 (2016) 3667-3676; Wei et al., Proc. Natl. Acad. Sci. USA 112 (2015) 2978-2983; Angew. Chem. Int. Ed. Engl. 53 (2014) 11822-11827.

564411 Point-of-Care Sepsis Diagnostics

Sepsis is a life-threatening organ dysfunction syndrome caused by a dysregulated and harmful host reaction to infection, accounting for at least 1/3 of all hospital deaths. A critical unmet need in controlling sepsis is the lack of quantitive, fast tests that produce actionable results in busy clinical settings. Here we aim to develop a new diagnostic platform for rapid, point-of-care (POC) sepsis detection. Termed IBS (integrated Biosensor for Sepsis), the approach leverages our breakthroughs: i) the newly-found pathophysiological role of cytokine interleukin-3 (IL-3) in sepsis, and ii) a hybrid magneto-electrochemical sensor for IL-3 detection. The developed diagnostic platform produces test results within 1 hour from native blood samples, and detects IL-3 at a sensitivity of <10 pg/mL; this performance is 5-times faster and 10-times more sensitive than a current gold standard enzyme-linked immusorbent assay. Using clinical samples, we show the potential of IL-3 as an early diagnostic biomarker. Compact and fast, the IBS platform can be readily integrated into clinical workflows, enabling timely diagnosis and proactive treatment of sepsis.

REFERENCE

Min, M. Nothing, B. Coble, H. Zheng, J. Park, H. Im, G. Weber, C.M. Castro, F.K. Swirski, R. Weissleder*, H. Lee*, “Integrated biosensor for rapid and point-of-care sepsis diagnosis,” *ACS Nano*, 12, 3378-3384 (2018). DOI: 10.1021/acsnano.7b08965 (Featured in MGH Blog and Nanowerk)

567640 Magnetic Blood Purification for Pathogen and Circulating Tumor Cell Removal

The magnetic separation of pathogenic compounds from body fluids is an appealing concept for both improved diagnosis and therapy with prospective applications in personalized medicine. Recently, removal of a diverse array of pathogens has been demonstrated by us and others using extracorporeal dialysis-type devices. Here, we report on the feasibility to remove pathogens and mammalian cells (e.g. circulating tumor cells) from body fluids in a continuous dialysis-type of setting. We employ magnetic nanoparticles functionalized with a newly developed antibody to capture and remove disease-causing factors from the blood.

We assess removal efficiencies from physiological fluids with or without white blood cells using a range of different magnetic bead sizes (50 nm – 4000 nm), concentrations and contact times. Both bacteria and tumor cells can be quantitatively removed from body fluids within acceptable times (1-2 minutes) and bead concentrations (0.2 mg per mL). Then, we exploit the opportunity for rapid and sensitive analysis of the captured entities on magnetic beads recovered from the magnetic separator to allow speedy identification and genetic profiling.

The present theranostic approach could significantly speed up detection and identification of disease-causing pathogens and at the same time provide an effective treatment modality by decreasing the pathogen load to bridge the time before appropriate (patient-specific) drugs can be administered.

1. I.K. Herrmann, M. Urner, F.M. Koehler, M. Hasler, B. Roth Z'Graggen, R.N. Grass, U. Ziegler, B. Beck-Schimmer, W.J. Stark, Blood Purification using Core/Shell Nanomagnets, Small 2010. Featured in Nature Nanotechnology, 5, 628 (2010).
2. M. Lattuada, Q. Ren, F. Zuber, M. Galli, N. Bohmer, M.T. Matter, A. Wichser, S. Bertazzo, G.B. Pier, I.K. Herrmann, Theranostic Body Fluid Cleansing: Rationally designed magnetic particles enable capturing and detection of bacterial pathogens, Journal of Materials Chemistry B, 2016.
3. N. Bohmer, N. Demarmels, E. Tsolaki, L. Gerken, K. Keevend, S. Bertazzo, M. Lattuada, I.K. Herrmann, Removal of Cells from Body Fluids by Magnetic Separation in Batch and Continuous Mode: Influence of Bead Size, Concentration and Contact Time, ACS Applied Materials and Interfaces, 2017.

568071 Design and Development of Inorganic Nanoparticles for Radioenhancement Therapy

Nanoparticle-based radio-enhancement has the prospect to improve cancer cell eradication by amplifying the damage caused by (X-ray) irradiation through ejection of secondary particles (electrons) leading to further oxidative stress. Gold nanoparticles are a natural choice because of their high atomic number and biological compatibility and are therefore the most researched nanoparticles. However, due to the heterogeneous results and the lack in mechanistic understanding, such nanoparticle systems have not yet entered clinics[1]. Most recently, nanoparticles based on more exotic hafnium dioxide have shown promising results in first clinical studies[2]. There is increasing evidence that radio-enhancement efficacy does not solely depend on high atomic numbers[3] but involves a complex cascade of secondary reactions.

Here, we will present a comprehensive nanoparticle-based radio-enhancement study including a selection of inorganic oxide nanoparticles (TiO_2, ZrO_2, HfO_2) with comparable morphologies, sizes and surface chemistries. We will report on uptake, subcellular distribution and radio-enhancement effects in radio-sensitive and resistant cells using clinically relevant exposure settings. The radio-enhancement efficacies are then benchmarked against the gold standards in the same exposure setting. We also show how the nanoparticles can be tailored to incorporate multimodal imaging possibilities[4]. Our comprehensive analysis will pave the way for a more rationalized nanoparticle design and development for nanoparticle-based radio-enhancement studies.

References

1. Ricketts, K.; Ahmad, R.; Beaton, L.; Cousins, B.; Critchley, K.; Davies, M.; Evans, S.; Fenuyi, I.; Gavriilidis, A.; Harmer, Q. J.; et al. Recommendations for Clinical Translation of Nanoparticle-Enhanced Radiotherapy. The British Journal of Radiology 2018, 91 (1092), 20180325.
2. Bonvalot, S.; Pechoux, C. L.; Baere, T. D.; Kantor, G.; Buy, X.; Stoeckle, E.; Terrier, P.; Sargos, P.; Coindre, J. M.; Lassau, N.; et al. First-in-Human Study Testing a New Radioenhancer Using Nanoparticles (NBTXR3) Activated by Radiation Therapy in Patients with Locally Advanced Soft Tissue Sarcomas. Clin Cancer Res 2017, 23 (4), 908–917.
3. Cui, L.; Her, S.; Borst, G. R.; Bristow, R. G.; Jaffray, D. A.; Allen, C. Radiosensitization by Gold Nanoparticles: Will They Ever Make It to the Clinic? Radiotherapy and Oncology 2017, 124 (3), 344–356.
4. Gerken, L. R. H.; Keevend, K.; Zhang, Y.; Starsich, F. H. L.; Eberhardt, C.; Panzarasa, G.; Matter, M. T.; Wichser, A.; Boss, A.; Neels, A.; et al. Lanthanide-Doped Hafnia Nanoparticles for Multimodal Theranostics: Tailoring the Physicochemical Properties and Interactions with Biological Entities. ACS Appl. Mater. Interfaces 2019, 11 (1), 437–448.

568572 PSMA-Targeted Glucose Oxidase Nanoparticles for Directed Cytotoxicity Toward Prostate Cancer Cells

Hormone-refractory prostate cancer is a deadly disease with few effective treatment options. Many types of cancer, including hormone-refractory prostate cancer, have shown susceptibility to glucose depletion and reactive oxygen species (ROS). The cytotoxic protein glucose oxidase (GOX) is of significant interest as an anticancer agent due to the ability of the enzyme to induce these effects of glucose depletion and ROS generation. Utilizing GOX as a cancer therapy, however, requires a safe and efficient delivery platform to limit GOX toxicity toward non-targeted cells while maximizing delivery to hormone-refractory prostate cancer cells. In this work, we aimed to enhance the cytotoxicity of GOX toward hormone-refractory prostate cancer cells by encapsulating the enzyme within nanoparticles permeable to glucose and target to prostate specific membrane antigen (PSMA). GOX was encapsulated within the cationic copolymer poly-L-lysine-grafted-polyethylene glycol (PLL-g-PEG) and functionalized with anti-PSMA monoclonal antibody. Characterization of the resulting nanoparticle complexes (anti-PSMA-PLL-g-PEG/GOX) indicated that they retained 80% of the encapsulated enzyme initial activity and were approximately 30 nm in diameter. PSMA-expressing LNCaP cells showed significant reductions in cell viability (~60%) following treatment with anti-PSMA-PLL-g-PEG/GOX, while negligible reductions in cell viability were observed for equivalent concentrations of unmodified GOX. Nanoparticle uptake and intracellular generation of ROS were observed in concurrence with the enhanced cytotoxicity of anti-PSMA-PLL-g-PEG/GOX toward LNCaP cells. These results suggest that the intracellular delivery of GOX is a useful approach to amplify enzyme cytotoxicity toward prostate cancer cells and could provide a basis for novel therapies against hormone-refractory prostate cancer.

577253 Hydrogel-Crosslinking By Minimally Invasive Surgical Modalities

In this study, an enzymatic crosslinked hydrogel formation is triggered using temperature sensitive liposomes (TSLs). The process is performed by encapsulating an enzyme inside TSLs and releasing it at the phase transition temperature (T_m) of the TSLs. At T_m, TSLs passes through a melting phase transition from gel to a liquid-crystalline phase. This transition increases membrane permeability which results in the release of the encapsulated molecules. Most studies with TSLs have applied superficial heating at body temperature which suffers from poor localization of heating and eventually poor localization of the release content. To generate the localized and controlled release of the enzyme at 44°C, focused electromagnetic hyperthermia or focused ultrasound can be used. An injectable probe having accessibility to the subcutaneous layer is used to inject a liquid polymer solution at the injured spot. The liquid contains enzyme encapsulated TSLs, suspended in the Hyaluronic acid- Tyramine conjugate (HA-Tyr) and hydrogen peroxide solution. Our focused hyperthermia or ultrasound system can solidify the polymer solution injected inside the body by releasing the crosslinkers from TSLs. By this way, nonspecific heating and crosslinking along with traditional invasive surgery can be minimized. This site-specific, soft hydrogel formation can achieve minimally invasive short-term implant for tissue engineering and many other clinical applications.

575266 Micromotors for Active Drug Delivery in Gastrointestinal Tract

Advances in bioinspired design principles and nanomaterials have led to tremendous progress in autonomously moving synthetic nano/micromotors with diverse functionalities in different environments. However, a significant gap remains in moving nano/micromotors from test tubes to living organisms for treating diseases with high efficacy. Here we present the first, to our knowledge, in vivo therapeutic micromotors application for active drug delivery to treat gastric bacterial infection in a mouse model using clarithromycin as a model antibiotic and Helicobacter pylori infection as a model disease. The propulsion of drug-loaded magnesium micromotors in gastric media enables effective antibiotic delivery, leading to significant bacteria burden reduction in the mouse stomach compared with passive drug carriers, with no apparent toxicity. Moreover, while the drug-loaded micromotors reach similar therapeutic efficacy as the positive control of free drug plus proton pump inhibitor, the micromotors can function without proton pump inhibitors because of their built-in proton depletion function associated with their locomotion. Overall, self-propelling microrobots can work as an active delivery technique that autonomously and precisely transports the therapeutic agents inside live animal's gastrointestinal tract, improving therapeutic efficacy for bacterial infection treatment. This technique opens the door for micro/nanorobots as an active delivery platform for medical treatment and is promising for a wide range of personalized diagnostic and therapeutic applications.

Refs:

Science Robotics, 2017, 2, eaam6431.

*Nature Communications***volume 8**, Article number: 272 (2017)

Angew. Chem. Int. Ed. 2017, 56, 2156 –2161.

ACS Nano 2016, 10 (10), 9536–9542.

ACS Nano, 2016, 10 (6), 5619–5634.

578272 Inflammation Targeted Intravenous Dendrimer-Drug Therapy for Age Related Macular Degeneration

Siva P. Kambhampati,[1,2] Jerry A. Lutty [2] and Kannan Rangaramanujam [1,2]

Inflammation mediated by activated microglia/macrophages (Mi/ma) is implicated in many ocular posterior segment diseases such as age-related macular degeneration (AMD), diabetic retinopathy (DR), glaucoma, retinitis pigmentosa (RP) and retinopathy of prematurity (ROP). Mi/ma under normal conditions perform immune surveillance and maintain retinal/choroidal hemostasis, but under disease conditions particularly in AMD, these cells migrate, and take part in intense cross-talk with other cells by secreting inflammatory cytokines, pro-angiogenic factors and influencing disease progression by causing oxidative stress and neovascularization. Current treatments for AMD involve frequent, localized intravitreal injections of anti-VEGF agents to neutralize the soluble VEGF but does not address the underlying inflammation. Further, these therapies doesn't work in 1/3 of the patients, and often associated with side-effects such as choriocapillaries atrophy, RPE loss and intraocular toxicity and there is no viable therapy for early AMD. Hence, targeting and delivering drugs to the key cells involved in disease progression will highly beneficial for both early and late stages, improve compliance and reduce costs. We have discovered that hydroxyl terminated PAMAM dendrimers when administered intravenously targets, localizes and retained in activated mi/ma, injured RPE and astrocytes in more than 5 ocular disease models [1,2]. We found that this unique intrinsic targeting of dendrimers is attributed to (1) the size, shape and neutral surface functionality of dendrimers, (2) Significantly reduced non-specific tissue interaction and its ability cross impaired blood retinal barrier, and (3) increased phagocytic activity of activated mi/ma [3,4].

In this study, we harnessed the targeting ability of dendrimers to deliver drugs (anti-inflammatory and anti-angiogenic) to the key cells (activated mi/ma and injured RPE) and evaluate is efficacy for early and late AMD in subretinal lipid injected rat AMD model. For early AMD, Dendrimer-N-Acetyl Cysteine (D-NAC) or dendrimer-triamcinolone acetonide (D-TA) monotherapy was utilized. When administered intravenously, D-NAC or D-TA demonstrated pathology dependent biodistribution by co-localizing and retaining in mi/ma and RPE cells for more than 21 days but cleared intact from systemic circulation and off-target organs within 24 hrs [2]. D-NAC or D-TA monotherapy resulted in attenuation for retinal and choroidal inflammation by suppressing inflammatory cytokine expression and promoting macrophage depletion in bleb areas compared to free drug controls. D-NAC or D-TA monotherapy resulted in significantly smaller CNV areas suggesting CNV suppression. D-NAC therapy resulted in phenotype switching of microglia from 'activated' to 'resting' thereby promoting restoration of retinal health. For late AMD, when a combination therapy of intravenous D-NAC + D-TA at day 10 onwards demonstrated CNV regression (~72%), reduction of oxidative stress, inflammatory and pathological angiogenic markers. Fluorescein angiography demonstrated that combination therapy resulted in reduced vascular leakage compared to free drug and untreated controls. Intravenous dendrimer-drug therapy did not demonstrate elevation of IOP or any signs of ocular and systemic toxicity [2].

Targeting and selectively co-localization of dendrimers, in activated mi/ma and RPE, makes them excellent carriers for systemic drug delivery to specific affected cells in the choroid/retina. Delivering drugs to the key pathological cells via dendrimers without any toxicity effects may provide opportunity to develop dendrimer-based therapies to achieve efficacy at significantly lower concentration with a more amenable and patient friendly delivery route particularly for dry AMD where there are no viable therapies available. The efficacy of D-NAC and D-TA in CNV suppression/regression, attenuation of retinal/choroidal inflammation may offer an effective treatment option for early and late stages of AMD and other ocular diseases where inflammation is involved.

[1] Center for Nanomedicine, Wilmer Eye Institute, Johns Hopkins School of Medicine, Baltimore, Maryland, USA.

[2] Department of ophthalmology, Wilmer Eye Institute, Johns Hopkins School of medicine, Baltimore,

Maryland, USA.

References:

1. Kambhampati, SP., Clunies-Ross, A.J., Bhutto, I., Mishra, MK., Edwards, M., McLeod, D.S., Kannan, RM., Lutty, GA. Systemic and Intravitreal Delivery of Dendrimers to Activated Microglia/Macrophage in Ischemia/Reperfusion Mouse. *IOVS*. 2015, 56(8), 4413-4424.
2. Kannan, RM., Lutty, G., Kambhampati, SP., Mishra, Mk., Bhutto. I.A. "Dendrimer compositions and their use in treatment of diseases of the eye." U.S. Patent Application 15/307,284, filed February 16, 2017.
3. Alnasser, Y., Kambhampati, SP., Nance, E., Rajbhandari, L., Shrestha, S., Venkatesan, A., Kannan RM., Kannan, S. Preferential and Increased Uptake of Hydroxyl-Terminated PAMAM Dendrimers by Activated Microglia in Rabbit Brain Mixed Glial Culture. *Molecules*. 2018, 23(5), 1025-1039.
4. Nance, E., Zhang, F., Mishra, M. K., Zhang, Z., Kambhampati, S. P., Kannan, R. M., & Kannan, S. Nanoscale effects in dendrimer-mediated targeting of neuroinflammation. *Biomaterials*. 2016, 101, 96-107.

580128 Micro and Nanotechnologies for Precision Medicine

Despite the plethora of new biotechnologies on the market, there is still a dearth of accurate, repeatable and reliable microtechnologies that can truly transform medical industry to date. Initially there were many attempts to simply miniaturize existing technologies, which led to many commercialization failures, since physics is not linear across spatial scales. Specifically, macroscale technologies hit fundamental size limits (sample size, abundance of chemical species, absorbance limits, etc), and thus inventing a micro- or nanofluidic technology often requires fundamental breakthroughs in physics, chemistry, fabrication, and materials engineering. Certainly, the last decade has brought about many of those new types of inventions, and in this talk I will highlight from an engineering perspective a few innovations from my laboratory at UCSB and/or companies I have founded that are relevant to the medical industry. In particular, I will highlight how fluids at the nanoscale behave different than micro- and macro scale counterparts because of the influence of the surface, and in particular, with electrical field driven systems, because of the influence of the electric double layer inherent at a solid-liquid interface of electrolyte-based analyte system (ie. biofluids). In addition, I will show how these differences can lead to unique physics couplings that allow for the detection of ions with selectivity and sensitivity that have never been yet achieved. Furthermore, this sort of detection leads to indirect detection of biomolecules, as ion concentration can change as biomolecular reactions occur. Finally, I will show results from the latest company I founded, that uses unique physics at the micro- and nanoscale to overcome the technological barriers needed to develop a painless, wearable glucose sensor that can be rapidly commercialized for patients with Type I diabetes, allowing them for more choices than the few that are out there.

579960 A Membrane Biochip Platform for Molecular Biomarker Diagnostics

Early diagnosis of cancer and cardiovascular diseases can be achieved by monitoring upstream gene expression pathways. Cells under acute stress form stress granules to protect their RNAs. Cells also gather and discard expired RNAs in nanoscale vesicles like liposomes, microvesicles or exosomes that can be secreted. As these nano-transporters prevent RNA degradation, their molecular cargo can induce disease states in other cells but can also incite collective immuno-response and defensive states. The latest liquid biopsy technologies hence profile circulating vesicular RNAs to detect irregular translation/expression. The diverse heterogeneity of nanocarriers and their molecular cargoes requires a systemic approach that is as complex as their genesis. My group integrates high-yield vesicle/molecule extraction, fractionation, purification and sensing electrokinetic modules into tunable high-throughput systems for large panels of pan-disease biomarkers. The modules contain electric field activated ion-selective media like gels and membranes with unique rectification, pH and ionic-strength control features. These ionic circuit components are integrated with a second layer of (detachable) electronic circuitry. We report some successful platform validation with spiked cancer cell media and clinical samples.

580331 Medical Applications of Microfluidics: From Blood Fractionation to Liquid Biopsy

The use of liquid biopsy for minimally invasive bioanalysis is revolutionizing healthcare, with particular promise in the diagnosis and treatment of cancer. Microfluidic devices based on inertial microfluidic approaches are receiving considerable attention for applications in liquid biopsy due to their label-free nature. However, the existing approaches to isolate, and analyze liquid biopsy samples are complex and deliver limited throughput, making them challenging to deploy as routinely-used tools. Further, many of these devices rely on sample dilution, which makes operation and deployment even more challenging. We are developing a microfluidic platform capable of label-free separation from unmodified whole blood samples to rapidly screen rare cell populations or fractionate blood cells. I will describe these efforts and provide a broader perspective on challenges and opportunities for microfluidic technologies in the areas of whole blood microfluidics for liquid biopsy and personalized medicine.

580279 Tracking Epitranscriptomics Modifications to Understand Early Health Effects of Oxidative-Prone Air

Ambient air pollution is associated with premature death caused by heart disease, stroke, chronic obstructive pulmonary disease (COPD) and lung cancer. However, the underlying mechanisms leading to detrimental effects on health by air pollution is yet to be understood. This has made it difficult the development of sensitive and specific biomarkers that, at the molecular level, can report cellular toxicity upon exposure to various air chemicals. Recent studies suggest that RNA oxidation is a sensitive environmental-related biomarker that is implicated in pathogenesis. In this work funded by the HEI through a Rosenblith Young Investigator Award, we applied a novel approach that integrates RNA-seq analysis with detection by immunoprecipitation techniques of the prominent RNA oxidative modification 8-oxo-7,8-dihydroguanosine (8-oxoG). Our goal was to uncover specific transcriptomics oxidation induced by VOCs/ozone mixtures in healthy epithelial lung cells that could be used as sensitive and early indicators of air-induced cell toxicity. We are investigating potential chemical modifications that are induced in mRNAs upon exposure to oxidative-prone air and to study how these perturbations affect normal patterns of cellular regulation. Specifically, we have been investigating: (1) what biochemical mechanisms lead to the breakdown of normal (healthy) transcriptome (mRNA) expression levels in various cellular pathways? And, (2) how is the breakdown of biological function at the molecular level caused by environmentally-induced chemical modifications? As such, we exposed BEAS-2B to the gas and particle phase products formed from reactions of 790 ppb acrolein and 670 ppb methacrolein with 4 ppm ozone (O3). Using this approach, we identified several potential direct targets of oxidation belonging to previously described pathways, as well as uncharacterized pathways on air pollution exposures. We demonstrated the effect of VOCs/O3 mixtures on the morphology of lung cells, suggesting the influence of selective mRNA oxidation in members of pathways regulating physical components of the cells as consistently observed with cellular phenotypes in COPD studies. In addition, we observed the influence of VOC/O3 mixtures on several specific metabolic pathways likely implicated with the incidence of mRNA oxidation and the deregulation of several proteins. Taken together, our study features the strength of combining traditional transcriptome analysis with epitranscriptome mechanistic characterizations to facilitate the discovery of underling processes that are not well described at the molecular level by current approaches. Thus, this presentation will focus on the ways that this research has informed the relationship between the atmospheric agents and molecular cellular responses implicated with diseases. We will discuss our findings uncovering specific mRNAs that are more susceptible to oxidative species and the impact of these modifications on overall cell function. Emphasis will be placed on how these studies have led to distinct biomarkers that can be used to sensitively assess toxicity effects.

568484 Tunneling Spectroscopy for Sequence and Structural Label Determination in Single DNA and RNA Molecules

In the push for precision medicine, nucleic acid sequencing is vital for compiling the big data libraries for genome characterization and in the clinic for point-of-care diagnostics. Nanoelectronic nucleic acid sequencing can provide an important alternative to traditional sequencing-by-synthesis by reducing sample preparation time, cost, and complexity as a high-throughput next-generation technique with accurate single-molecule identification. However, sample noise and signature overlap due to varying nucleotide conformations continue to prevent high-resolution and accurate sequencing results. We have developed a nanoelectronic method to combat these issues that combines non-perturbative quantum tunneling spectroscopy with machine learning classification algorithms to distinguish the nucleotides adenine (A), guanine (G), cytosine (C), thymine (T), and uracil (U), as well as nucleotide chemical labels. These results from our quantum molecular sequencing (QMSeq) approach not only present a promising path forward for single-molecule DNA sequencing [1], but also a method for single-molecule direct sequencing and structural characterization of RNA [2].

Using tunneling spectroscopy measurements, we probed the molecular orbitals of chemically distinct nucleobases within DNA and RNA macromolecules immobilized with restricted conformational freedom on a chemically-modified surface. From these measurements, theoretical models for quantum tunneling were combined with transition voltage spectroscopy to obtain twelve biophysical parameters unique to the nucleotides within the electronic tunneling junction. The twelve parameters serve as a comprehensive molecular fingerprint for the nucleotides, facilitating their discrimination and identification of structure-dependent chemical labels through machine learning. We show a high accuracy for both nucleotide discrimination (>99.8%) and chemical label identification (>98%) with a relatively modest molecular coverage (35 repeat measurements).

These results have significant implications for the development of robust and accurate high-throughput nanoelectronic DNA and RNA sequencing techniques. Additionally for RNA, we have shown the potential for simultaneous sequencing and structural mapping of single unknown RNA molecules, paving the way for probing the sequence-structure-function relationship within the transcriptome at an unprecedented level of detail.

[1] Korshoj, Afsari, Khan, Chatterjee, Nagpal, *Small* 13 (11), 1603033 (2017).

[2] Abel, Jr., Korshoj, Otoupal, Chatterjee, Nagpal, *Chemical Science* 10 (4), 1052-1063 (2019).

570394 Profiling Cell-Type-Specific Epigenomic Changes Associated with BRCA1 Mutation in Breast Tissues Using a Low-Input Microfluidic Technology

Previous studies have shown that genes can be switched on or off by age, environmental factors, diseases, and lifestyles. The open or compact structures of chromatin is a crucial factor that affects gene expression. Epigenetics refers to hereditary mechanisms that change gene expression and regulations without changing DNA sequences. Epigenetics plays critical roles in cell differentiation and disease processes. Breast cancer is the second mortal cancer among American women, following lung cancer. BRCA1 and BRCA2 are extremely influential in repairing DNA double-strand break through homologous recombination and suppressing DNA replication stress. Gene mutations in BRCA1 and BRCA2 increase the risk of breast cancer. Human female mammary epithelium is a two-layer tissue, including the inner (luminal) layer and outer (basal) layer. The luminal progenitors (LP) and mature luminal cells (ML) are from the inner layer and basal cells (B) are from the outer layer. The fourth cell type is stromal cells (S) from mesodermal tissue. Recent advances in epigenetics enable us to use next-generation sequencing to elucidate epigenomic dynamics during disease and development. Conventional chromatin immunoprecipitation followed by sequencing (ChIP-seq) necessitates the use of up to 1 - 10 million cells. Using microfluidic-oscillatory-washing-based ChIP-Seq (MOWChIP-Seq) developed in our lab, we profile epigenomes using as few as 100 cells as starting material. We used MOWChIP technology to examine four cell types from breast tissues of cancer-free patients, including normal (non-carrier) and BRCA1 mutant (carrier).

Cancer-free BRCA1 mutant (carrier) or normal (non-carrier) breast tissues were obtained from adult female patients who underwent cosmetic reduction of mammoplasty, diagnostic biopsies, or mastectomy. Four different cell types from the same patient were sorted by FACS. The coated beads and sheared chromatin (from 50k cells) were loaded to MOWChIP device for ChIP process. We examined genome-wide H3K4me3 and H3K27ac. After ChIP, DNA was purified, used for library preparation and sequenced by Illumina HiSeq 4000. The digital data were analyzed with Bowtie, MACS, and DiffBind.

Using the MOWChIP device enables us to perform ChIP on rare samples in less time and with improved quality of sequencing data. Stromal cells displayed the most differences in H3K27ac peak signal intensity between carrier and non-carrier groups, with the vast majority of peak regions having higher H3K27ac levels in non-carriers. Similarly, non-carrier mature luminal cells also had increased H3K27ac peak signal intensity, though very few peaks were significantly different between the two groups. In contrast, non-carriers had lower H3K27ac peak signal intensity across thousands of regions in both basal and luminal progenitor cells. These differences support widespread and cell-specific epigenomic changes associated with the BRCA1 mutation. Our work suggests strong genetic and epigenetic interactions due to mutations.

575974 A Metabolomics Approach to Predict Chemotherapy-Induced Peripheral Neuropathy

Introduction: Vincristine is a core chemotherapeutic drug administered to pediatric Acute Lymphoblastic Leukemia (ALL) patients. Although it is effective and has been in use for more than 50 years now, it has a dose-limiting toxicity: vincristine-induced peripheral neuropathy (VIPN). VIPN is characterized by symptoms such as a painful, numbing and tingling sensation felt in the palm and feet, with potential long-term effects[1]. Predictors and mechanism of VIPN incidence remain unclear, especially during the early phase of the treatment. Historically, an empirical cap of 2 mg is set for vincristine dosage. Due to this, a cohort of patients not susceptible to VIPN receives sub-therapeutic treatment, while another cohort may experience severe neuropathy. Thus, it is of interest to find predictors/biomarkers which can discriminate between these cohorts based on VIPN severity and hence aid in treatment personalization.

Methods and Results: Here, we employed a metabolomics approach to find metabolites as biomarkers that can accurately predict VIPN in pediatric ALL patients. We performed untargeted metabolite profiling using tandem mass spectrometry on blood samples of ALL patients undergoing chemotherapy with vincristine as the primary drug. We analyzed the samples at three time points: day 8, day 29 and around 6 months during the treatment. We classified these patients as being susceptible to high or low neuropathy based on total neuropathy score calculated frequently throughout the course of the treatment. Following metabolite profiling, we used machine learning techniques to analyze the profile data and determine a small set of metabolites highly predictive for overall VIPN susceptibility at those time points. Firstly, we performed a principal component analysis which showed that the metabolite profiles cluster according to the time point, implying that they need to be analyzed independently, rather than longitudinally. Then, we used support vector classifier along with recursive feature elimination algorithm to find the small sets of predictive metabolites. Cross-validation models built with these selected metabolites as predictors had an area under receiver operating characteristics (AUROC) curve of approximately 0.95 at all the time points. Furthermore, we performed rigorous manual validation of the chosen metabolites' chromatogram peaks to ensure that they were integrated correctly by the mass spectrometry instrument. After manual validation, we selected the correctly integrated metabolites and built predictive models with them. Models built using the day 8 and 6 months data were more accurate, with AUROC greater than 0.9 in both the cases. Thus, one could use these trained models at day 8 and 6 months during the treatment to predict VIPN susceptibility and subsequently adjust vincristine dosage. We further identified few of these metabolites using their mass, retention time and adduct information. Lastly, we found relevant biological pathways associated with them.

Conclusion: We found small sets of metabolites that could accurately predict VIPN in pediatric ALL patients. We then built predictive models with these metabolites which can potentially aid in vincristine dosage decision making for clinicians, and subsequently improve treatment efficacy for patients. This approach can further be applied to predict peripheral neuropathy as a result of other chemotherapeutic drugs as well.

References:

1. Gidding, C. E. M., et al. "Vincristine revisited." Critical reviews in oncology/hematology 29.3 (1999): 267-287.

576560 An Exposome Connectivity Paradigm for the Mechanistic Assessment of the Effects of Prenatal and Early-Life Exposure to Metals on Neurodevelopment

The introduction of exposome studies to research started to make clear that the timing of exposure to environmental factors could be even more important than the dose. Given the complexity of the individual lifetime exposome assessment, a broad array of technologies must be employed. The connectivity approach to environmental health, builds upon the connectivity across different biological scales in a systems biology approach to elucidate the mechanisms underlying the environmental burden of disease. Adopting a data-driven paradigm that is guided by systems biology principles this approach couples comprehensiveness in exposome and health associations and biological plausibility. Thus, it can unravel causal links between environmental exposures, genetic predisposition and early biological markers of adverse health outcomes without needing to revert to very large population sizes. In the frame of HEALS (Health and Environment-wide associations via Large population Studies) we have been applying the exposome connectivity paradigm on several pre-existing cohorts re-analysing biobanked samples using a full multi-omics platform coupling epigenetics, SNPs, metabolomics and adductomics, and applying integrative bioinformatics and exposome-wide association algorithms to draw links between combined exposures to metals and endocrine disrupters and metabolic pathway dysregulation, as well as between metabolic pathway perturbations and clinically observed phenotypes of neurodevelopmental disorders such as problems in linguistic, motor development and cognitive capacity.

The current study, called PHIME, presents an integrated methodology about the investigation of the effects of prenatal and postnatal exposure to metals on child neurodevelopment. Heavy metals are well known neurotoxicants and exposure to heavy metals has been extensively associated with neurodevelopmental disorders. The links between in utero exposure to metals, metabolic pathway deregulation, and clinically observed phenotypes of neurodevelopmental disorders were drawn through a urinary and plasma untargeted metabolomics analysis using both nuclear magnetic resonance (NMR) spectroscopy and liquid chromatography-mass spectrometry (MS), followed by integrative bioinformatics and exposome-wide association algorithms. Heavy metals prenatal exposure was determined by measuring mercury in hair samples collected at birth, while cord blood and breast milk samples were analysed for mercury, cadmium, lead, and arsenic, as well as for essential elements (selenium, zinc, copper). Additional factors considered as exposure and effects modifiers included in the study were sociodemographic and anthropometric parameters, as well dietary factors. Cognitive function, language, fine and gross motor development were assessed in children at the age of 18 months by the Bayley Scale for Infant Development (Bayley-III) development tool. The Wechsler Intelligence Scale for Children - Fourth Edition was used for the assessment of the Full-Scale Intelligence Quotient (FSIQ). Finally, individual metabolome profiles have been derived analyzing biobanked urine and plasma samples taken at birth, and urine and plasma samples taken from children at the age of 8, using two different analytical techniques, LC-MS and NMR. These metabolic profiles were used for metabolic pathway analysis at the individual level and the identified perturbed pathways were associated with both exposure and outcome parameters using EWAS.

In order to generate a list of candidate biomarkers for the stressors of interest, which were identified by untargeted analysis of metabolic profile, a literature search was performed including combinations of the following terms: ''exposure'', ''metals'', ''neurodevelopment disorders'', ''biomarkers'' and ''underlying mechanisms''. The query was perform for both levels of analysis (molecular biomarkers and pathway analysis) for all the experiments. It was designed not only to determine candidate biomarkers based on their known biological function, but also to avoid the exclusion of a biomarker, which is associated with the occurrence of a stressor, even though there is yet paucity of knowledge regarding its exact biological function.

The imbalance between the cellular reactive oxygen species (ROS), which may be an effect of exposure to metals, and the inability of the cell to detoxify them, leads to oxidative stress. According to literature, abnormalities in the citric acid cycle, urea cycle, and amino acid metabolism play a key role in the pathogenesis of oxidative stress. Formate, 2-oxoglutarate, isocitrate, glycerol, carnitine, glutathione, methionine, cysteine, pyruvate, N-acetylglutamic acid, β-alanine, serine, arginine, citrulline, tryptophan, alpha-D-glucose, (S)-lactate and acetate which have been detected in samples from PHIME cohort study, could play the role of candidate biomarkers for the neurodevelopmental disorders due to oxidative stress. In addition, the perturbations of the identified pathways, for the homeostatic operation of which the presence of the above biomarkers is crucial must be examined as a putative underlying mechanism. Some of these identified pathways are: S-methyl-5-thio-Alpha-D-ribose 1-phosphate degradation, folate metabolism, serotonin degradation, taurine biosynthesis, citrulline-nitric oxide cycle etc. Henceforth, dysfunctions in carnitine metabolism may affect calcium homeostasis, which is involved in the oxidation of phosphorylation, leading to neurodevelopmental disorders. Biochemical markers directly or indirectly related to mitochondrial dysfunction, found to participate in dysregulated metabolic pathways were the carnitine, alanine, lactate, pyruvate, lysine and the acylcarnitine.

According to EWAS analysis, the concentration of lead (Pb) found in mother's plasma was negatively related to the metabolic pathways related to the effects of nitric oxide and the differentiation pathway. Cadmium (Cd) concentration levels showed a negative correlation with the degradation (III) of melatonin and dopamine, and the presence of mercury (Hg) with the metabolic pathway tyrosine degradation I. Selenium (Se) concentration (Se) found to be negatively correlated with the tetrapyrrole biosynthesis II and heme biosynthesis pathways, while copper (Cu) with caffeine and theobromine metabolism. Finally, manganese (Mn) appears to adversely affect the metabolic pathways of fatty acid biosynthesis, glycolysis/gluconeogenesis, amino sugar and nucleotide sugar metabolism, porphyrin and chlorophyll metabolism. With respect to children's samples, postnatal exposure to cadmium (Cd) is negatively related to the pathways of fatty acid biosynthesis, amino sugar and nucleotide sugar metabolism, porphyrin and chlorophyll metabolism and glycolysis/gluconeogenesis. Selenium (Se) levels was negatively associated with dopamine degradation.

Besides the negative association of the aforementioned pathways with neurodevelopment, EWAS analysis indicated that nutrition, as well as social factors, play an important role. In particular, the frequency of fish consumption during the pregnancy and the first 8 years of life, and the frequency of mushrooms consumption by children showed negative correlation with neurodevelopment. Other factors negatively correlated with neurodevelopment were the partner's age, maternal weight, and the source of drinking water.

From the cohort participants, one child has been diagnosed with Autism spectrum disorder (ASD). The exposure levels to As and Hg were above the 95th percentile, and according to untargeted metabolomics analysis, the levels of L-arginine, and L-tryptophan were above the 95th percentile, while the (S)-Lactate was below 5th percentile. The aforementioned results suggest major disturbances to cells biochemistry, which resulted in the impairment of antioxidant defense mechanisms leading to the clinically observed results in linguistic, motor development and cognitive capacity.

Two are the main characteristics that stand out from this study. The fact that we have chosen to consider factors such as dietary habits or socio-economic status as additional parameters associated with child neurodevelopment. Moreover, we were particularly interested in capturing the overall metabolome perturbation and associate it with (i) the totality of environmental exposure factors we could quantify; and (ii) clinically observed neurodevelopmental disorders. The scope of the study was to investigate how the perinatal and early-life exposome affected child neurodevelopment. Exposome-based associations of clinical or sub-clinical health outcomes with the early-life external and internal exposome supports the elucidation of the mechanisms through which xenobiotics interact with and eventually perturb cell metabolism to induce specific pathways of toxicity in infants and young children. However, in vitro testing coupled to targeted metabolomics on metabolically active relevant cell lines could be a viable way forward towards providing the mechanistic evidence of the observed exposome-wide associations.

578313 A Systems Approach to Understanding Drug Response in a Heterogeneous Tumor-Cell Population

Single-cell heterogeneity is pervasive across multiple molecular levels within a tumor and supports multiple mechanisms through which cellular subpopulations that are inherently drug resistant can arise or can acquire resistance during treatment. These issues hinder our ability to develop effective treatment strategies. A quintessential example of tumor cell heterogeneity is Glioblastoma (GBM), a highly aggressive and lethal form of primary brain cancer. One approach to address GBM cellular heterogeneity has been to apply high-throughput drug screens in order to identify novel single or combination drug therapies that inhibit growth of glioma stem-like cells (GSCs), a clinically relevant subpopulation of tumor cells that drive tumor formation/recurrence and are resistant to the current standard of care (surgical resection and treatment with the chemotherapeutic temozolomide). Results from our collaborators have demonstrated remarkable differences in the response of patient-derived GSCs to the drug pitavastatin, which has shown potential to inhibit GSC growth. These differences are due in part to the heterogeneous population structure of GSCs, which contain stem-like cells that differ in their tumor-initiation ability, molecular signatures, and drug responses. Understanding how a tumor-cell population is structured (i.e. proportions of subpopulations) and the regulatory mechanisms (e.g. transcription factor and miRNA regulators) that relate or distinguish these subpopulations would provide deeper insight into how tumor-cell heterogeneity contributes to overall tumor-cell population drug response. In this work, we apply a systems approach to determine the regulatory mechanisms that *relate or distinguish* cellular subpopulations by analyzing the transcriptomic responses of patient-derived GSCs and their time-course response to pitavastatin. Using network inference methodologies, we have identified several transcriptional programs (i.e., transcription factors and corresponding co-regulated target genes) that underlie these distinct subpopulations. This analysis provides insight into potential transcription factors that can be targeted to modulate the drug-response of subpopulations that are resistant to pitavastatin. More generally, this approach will inform the rational selection of molecular targets to attack specific drug-resistant subpopulations within a tumor.

575352 Quantitative Label-Free Interference-Based Phenotyping of Cancer Cells for Liquid Biopsy Applications

Reflection interference contrast microscopy (RICM) is a label-free imaging technique that provides a unique non-invasive "view from below" perspective of micro-scale objects interacting with a surface. RICM's interferograms embed precise nanometer-scale topographical information about the object under observation, making this microinterferometric approach ideal for phenotypic screening at the single-cell level based on cell adhesion phenomena. In this scenario, employing a large illumination numerical aperture (INA) becomes essential to enhance the contrast of the cell membrane interacting with the surface and produce high quality interferograms. However, important challenges emerge when trying to perform quantitative analysis of such RICM images, due to the lack of sophisticated RICM models that properly interpret intensities as membrane-surface separation distances when using large INAs. Here we present advances that enable quantitative RICM analysis over the full range of possible INAs. We introduce a new model of RICM image formation that successfully accounts for polarization effects, in both liquid and air environments, while facilitating the formulation of an absolute intensity scale. These fundamental advances not only improve quantification of membrane-surface separation but also provide a platform for consistent quantification and characterization of the morphology of the adhesion patch. Such unparalleled capabilities are used here to perform quantitative label-free phenotyping of cancer cells of low (LNCaP) and high (PC3) metastatic potential as they dynamically interact with a glass surface. Our findings illustrate the great potential of this approach for label-free identification of circulating tumor cells (CTCs) in liquid biopsy applications.

575863 Rapid Expansion of Cytotoxic T Lymphocytes in a Centrifugal Bioreactor with Applications in Cancer Immunotherapy

Adoptive cell therapy (ACT) has proven to be a promising form of cancer treatment; however, it lacks accessibility due to current expansion techniques. ACT requires rapid expansion of high-density cell cultures *in vitro* prior to patient administration, and depending on the stage of cancer, the patient may be required to go through multiple treatments. To address this high cell demand, researchers at Washington State University have developed a novel centrifugal bioreactor (CBR), which has been utilized for a number of applications. The CBR operates in perfusion at high centrifugal forces, while the cells are continuously fed with fresh medium. The balance of forces along with nutrient maintenance has allowed us to reach densities of 10^8 cells/mL. To further study the CBR for immunotherapy purposes, we are rapidly expanding cytotoxic T lymphocytes (CTLs) from a bovine system; cattle produce a similar immune response to that of cancer patients when affected by the *Theileria parva* parasite. In order to optimize cell growth in the reactor, we are producing a kinetic model to understand how the cells respond to environmental changes when consuming nutrients and producing toxic metabolites such as ammonium ion and lactic acid. Another point of interest for this project is in utilizing some form of biotin-streptavidin-based anti-CD3 support to enhance cell activation and proliferation before and during cultivation in the CBR. The results of these studies will be presented at the annual AIChE meeting.

575494 Enhancing Mass Transfer of Vitamin E to Inflamed Bovine Articular Chondrocytes through Perfusion

Articular cartilage is an avascular tissue with diffusion limited nutrient and oxygen transfer, making its ability to self-heal and regenerate limited. Osteoarthritis (OA) was traditionally defined solely as the degradation of cartilage and was not considered an inflammatory disease. However, the inflammation of the synovial membrane has been observed in a significant portion of OA patients. In addition, several recent studies have proved the presence of inflammatory markers, including interleukins, nitric oxides and macrophage inflammatory proteins, in serum and synovial fluid of OA joints. These findings transformed how researches define and develop treatments for OA.

Nutraceuticals are food components that have medicinal benefits in addition to their nutritional value. They can reduce inflammation by blocking the expression of iterluken-1, downregulating the overexpression of cell adhesion molecules and by their anti-oxidative characteristic, they scavenge reactive oxygen species (ROS) and free radicals.

In this study, inflammation is induced in bovine articular chondrocytes (bAChs) by the addition of IL-1ß. They are then cultured in micromass and bioreactor cultures and supplied with a nutraceutical (vitamin E) containing growth medium. A centrifugal bioreactor (CBR) is used to grow cartilage tissue under perfusion. We hypothesize that perfusion will enhance the mass transfer of nutraceuticals to the grown cartilage tissue, reflected by a significant decrease of nitric oxide content in the culture medium and the gene expression of inflammatory markers like TNF-a and NOS2 in comparison to micromass cultures.

576017 A Rapid Convection PCR-Based Diagnostic Platform

Technologies based on the polymerase chain reaction (PCR) are a gold standard for laboratory-based diagnostics. And these capabilities are enhanced through the use of multiplex reaction and target detection approaches, enabling highly specific analysis to be performed. But deployment of PCR in field-based or point-of-need settings has been limited by difficulties in adapting this instrumentation to a portable format. Here we describe the development of an analytical platform that exploits thermal convection to perform rapid PCR thermocycling in a manner amenable to operation in settings with little or no laboratory infrastructure. First, we establish a link between the PCR tube design (which governs the physics of convective flow and thermocycling) and the primer design (which regulates the PCR biochemistry). We systematically characterize this previously unexplored interplay, making it possible to identify optimal conditions that favor rapid replication of the desired target. We then extend these insights to introduce a flow-dependent correlation between replication efficiency and target GC content. Finally, we combine these insights with in-situ melting curve detection to enable robust, sensitive PCR-based analysis in a format that is readily deployable at the point-of-need.

560136 Digital Single Cell Profiling for Point-of-Care Cancer Diagnosis

The global burden of cancer, severe diagnostic bottlenecks in underserved regions, and underfunded health care systems are fueling the need for inexpensive, rapid and treatment-informative diagnostics. Based on advances in computational optics and deep learning, we have developed a low-cost digital system for breast cancer diagnosis of fine needle aspirates. Here, we show high accuracy in classifying breast cancer types using deep-learning based analysis of sample aspirates. The image algorithm is fast, enabling cellular analyses at high throughput, and the unsupervised processing allows use by lower skill health care workers. The system could be further developed for other cancers and thus find widespread use in resource limited settings to improve global health.

Reference

[1] Min J†, Im H†, Allen M, McFarland PJ, Degani I, Yu H, Normandin E, Pathania D, Patel JM, Castro CM, Weissleder R*, Lee H*, Computational optics enables breast cancer profiling in point-of-care settings, ACS Nano, 12(9), 9081-9090 (2018).

[2] Kim S, Wang C, Zhao B, Im H, Min J, Choi H, Tadros J, Choi N, Castro CM, Weissleder R, Lee H*, Lee K*, Deep transfer learning-based hologram classification for molecular diagnostics, Scientific Reports, 8:17003 (2018).

Validation of a New Computational Fluid Dynamics Model to Predict Turbulent Flow Damage for the US FDA Critical Path Initiative Centrifugal Blood Pump

Mesude Ozturk[a], Edgar A. O'Rear[c,d], and Dimitrios V. Papavassiliou*[b]

[a]Department of Chemical Engineering, Cumhuriyet University, Sivas, TURKEY
[b]Department of Chemical, Biological and Materials Engineering, University of Oklahoma, Norman, OK, USA
[c]University of Oklahoma Bioengineering Center

(*) Dimitrios V. Papavassiliou

Address: School of Chemical, Biological, and Materials Engineering

Bioengineering Center

University of Oklahoma

Sarkeys Energy Center, Room T-301

100 E. Boyd St.

Norman, OK 73019

Phone: (405) 325-0574

Fax: (405) 325-5813

Email: dvpapava@ou.edu

ABSTRACT

Blood constitutes about 8% of an average adult's body weight and its role is to transport material to and from tissue, prevent fluid loss, and defend the body. While plasma consists of water, protein, inorganic salts, and organic substances, blood cells include white blood cells, platelets, and red blood cells (erythrocytes). For every 600 red blood cells (RBC), there are around 40 platelets and 1 white cell, while the human body contains approximately 25 trillion erythrocytes. Red blood cells circulate in bloodstream and transport oxygen approximately for 120 days then broken down mechanically or biologically. However, the normal life span of blood components can be disturbed by prosthetic heart devices which cause non-physiological conditions, such as turbulent blood flow. Turbulent blood flow in artificial hearts and ventricular assist devices (VAD) cause hemolytic as well as sublethal red blood cell (RBC) damage, a major concern when designing prosthetic heart devices. In this work, the flow field through a centrifugal blood pump that has been identified as a case study through the Critical Path Initiative of the U.S. Food and Drug Administration (FDA) [1] was simulated under their specified conditions of operation and turbulence effect on RBC damage was investigated. Hemolysis was predicted using a validated computational model [2] which uses the Kolmogorov length scale (KLS) as an indicator of the degree of turbulence. Predictions are also to be compared to the other experimental findings on release of a centrifugal blood pump.

Key Terms: Computational Fluid Dynamics, Red blood cell trauma, Hemolysis, FDA, Centrifugal blood pump, Turbulence, Kolmogorov length scale, erythrocyte, artificial organs

1. INTRODUCTION

Ventricular assist devices (VAD) offer an effective therapy to the patients who are at an advanced level of heart failure waiting for a heart donor or who are not suitable for transplantation. [1] Turbulence is often a component of flow through ventricular assist devices. Therefore, examining the turbulence effects on red blood cell (RBC) trauma is of major importance in the design and development of VADs. It has been shown that hemolysis can be related to the surface area of eddies with sufficient energy dissipation to cause cell damage. [2-4] Consequently, hemolysis can be examined based on an extensive property (eddy surface area) of energy intensity for specific KLS (Kolmogorov Length Scale) throughout the domain of turbulent flows. [2-4]

Computational Fluid Dynamics (CFD) has been widely used to predict hemolysis as a function of flow field conditions found in cardiovascular devices and has often been used to analyze, improve, and optimize VADs. [5-10] CFD is also essential in the development process of rotary blood pumps such as evaluation of potential blood trauma, characterization of hydrodynamic performance and optimization of pump design. [11-14] Correct measurements of flow parameters, such as the instantaneous spatial distribution of the rate of dissipation of turbulent kinetic energy (ε), are difficult in the laboratory, which makes CFD essential for these systems. Furthermore, several different implant designs can be created in silico and examined for a broad range of operating conditions without the time and expense necessary for the production and testing of prototype devices.

Since invitro blood tests are demanding, animal tests are challenging and expensive, and flow in medical devices is completely associated to blood damage; the United States Food and Drug Administration (FDA) launched a Critical Path Initiative (CPI) project to standardize and promote the use of computational fluid dynamics (CFD) for the design and optimization of blood-contacting devices. Two standard flow models for medical devices using CFD was studied. The first study was initiated in 2008 to evaluate CFD as a biomedical research tool using a simple nozzle model. The second study intended to assess CFD predictions and limitations in characterizing flow and predicting blood damage in a generic centrifugal blood pump. [15-17]

In the present work, we perform hemolysis estimations for the flow field through the pump at different flow conditions and pump rotation rates by using a new model [4] that takes into account the complexity of turbulence by giving varying weight to eddies of different sizes and by considering the extent of those eddies throughout the device. Examining hemolysis as a function of the eddy surface area provides a relation with an extensive property of the flow – an approach that is more appropriate to account for the complexity of flow in medical devices. A small, highly turbulent region in the device may contribute significantly to hemolysis, and the size of such regions is also important to the extent of damage. [2-4]

2. METHODS

2.1 Blood Pump Model

The geometry for the simplified centrifugal blood pump model shown in Fig. 1 was created using computer-aided design as part of the FDA's CPI CFD and Blood Damage project. [15] The pump has a design that allows operations of wide range of pressure and flow conditions. The acrylic rotor of pump is 5.2 cm in diameter, the four filleted blades of rotor is in 3 mm tall and 3 mm wide. The blades are attached to a 3.2 mm diameter stainless steel shaft and orthogonally positioned on a 4 mm thick rotor base. PIV and hemolysis tests are performed in presence of spacers which are used to have 1 mm clearance gaps between the housings and the front and back of the rotor.

The flow domain is meshed in ANSYS® ICEM CFD® (v. 14.0) using the robust Octree method and is optimized based on the orthogonal quality of the mesh. Solutions were obtained using ANSYS® FLUENT® (v. 14.0) with second order upwind spatial discretization and first order temporal discretization. A grid independence analysis for convergence of pressure and velocity predictions indicated a grid consisting of approximately 2 million cells and 700,000 nodes was sufficient for capturing the characteristics of the flow field. Experimentally, flow through the pump is time-dependent due to the movement of the rotor and, as such, the flow was modeled as unsteady. The details of the model and the flow conditions can be found in the literature[18].

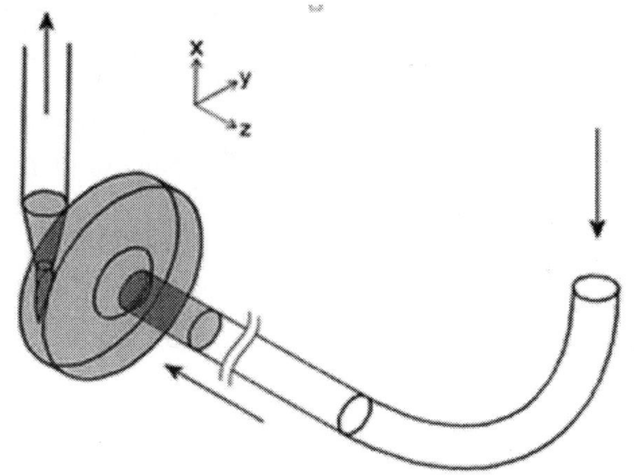

Figure 1: llustration of the FDA's simplified centrifugal blood pump model. [15]

2.2 Computational Approach

Simulations were performed at conditions specified in the CPI with rotor speeds of 2500 rpm, an operating temperature of 23°C, a flow rate of 2.5 lpm, and with 36% hematocrit. Flow was modeled as turbulent and Newtonian. The Shear-Stress Transport k-ω (SST k-ω) model was used for turbulent flow simulations. The SST k-ω model is accepted to be more suitable for internal flows with a low Reynolds number (Re < 10,000) and hence for calculating turbulence in the flow in heart valve. [19, 20] A sliding mesh approach was used with 200 time steps per rotation and a time step size on the order of 0.1 msec. A maximum of 30 iterations per time step was allowed for the solution to reach the minimum convergence criteria of 10^{-4} for all residuals. Flow field solutions were allowed to develop for at least 1000 msec (i.e., 50 complete rotations) before data sampling and particle injection were initiated.

In order to economize the computational calculations, we took advantage of rotor symmetry, a quarter of a full rotation was taken for the simulations and then divided by 10 to give a sequence of angular displacements (i.e., 10 different blade positions).

Several planes were created in the rotor from bottom to top and KLS values are calculated for each plane and for each of the different blade positions. Therefore, each plane has a set of 10 different KLS spatial distributions. KLS size distributions of each position were calculated using eddy analysis [2-4] and plotted.

Hemolysis values for the 10 angular displacements of each plane were calculated by using Equation 1. [4]

$$HI = 1.62*10^{-7} + 1.82*10^{-7} * t + 3.08*10^{-5} * EA_{KLS(0-3)} + 3.42*10^{-6} * EA_{KLS(4-6)} + 1.72*10^{-6} * EA_{KLS(7-9)} \quad (1)$$

where HI: hemolysis index (%), t: exposure time, and $EA_{KLS(R1-R2)}$: cumulative eddy surface area per volume for KLS size of R1 to R2 μm KLS. Here volume refers to the turbulent region demarcated by 10 μm KLS iso-turbulence lines.
The equation above was applied to predict single pass hemolysis in the pump for the different angular displacements. Hemolysis values were summed to get the total hemolysis prediction of the 10 rotational positions. Then the summed hemolysis value was multiplied by 4 to account for a full rotation of the rotor. Since hemolysis is calculated based only on the volume with KLS values of10 μm or less, the hemolysis value must be corrected for the full volume of the pump. Total hemolysis is diluted by the total volume and presented here. These steps were repeated for different plane separations to see the hemolysis changes.

4. RESULTS AND DISCUSSION

4.1. Calculation of Eddy Size Distribution and Kolmogorov Length Scale

The KLS values are examined thoroughly in the rotor where hemolysis takes place. Each plane of each rotation has a different range of KLS values. All the results are not shown here for simplicity. For example, there are more than 100 different KLS ranges for 2500 rpm rotor speed as can be seen on Table 1.

Table 1. KLS values on planes for 2500 rpm rotor speed.

Planes Plane_	8260 (8310) KLS µm	8265 KLS µm	8270 KLS µm	8275 KLS µm	8280 KLS µm	8285 KLS µm	8290 KLS µm	8295 KLS µm	8300 KLS µm	8305 KLS µm
0.001	14.4 - 892.3	14.6 - 898.1	15.6 - 890.8	16.3 - 882.1	16.4 - 889.0	16.5 - 880.7	16.3 - 873.4	16.9 - 864.0	16.8 - 851.2	17.0 - 849.8
0.002	15.9 - 232.6	16.0 - 231.9	16.0 - 231.5	15.5 - 232.2	15.5 - 233.1	15.6 - 233.7	15.7 - 234.1	15.8 - 234.2	16.0 - 234.1	16.4 - 233.7
0.003	15.2 - 238.5	14.6 - 239.4	14.6 - 240.2	14.7 - 240.9	14.9 - 241.2	15.1 - 241.4	15.3 - 241.2	15.4 - 240.8	15.5 - 240.2	15.6 - 239.5
0.004	9.3 - 466.7	9.0 - 466.7	8.9 - 462.6	9.0 - 455.9	9.1 - 452.8	9.2 - 458.3	9.2 - 464.1	9.4 - 466.3	9.6 - 462.4	9.8 - 462.6
0.005	11.1 - 315.4	10.6 - 315.5	10.6 - 315.8	10.7 - 316.1	10.9 - 316.3	11.1 - 316.3	11.4 - 316.3	11.6 - 316.0	11.7 - 315.8	11.9 - 316.5
0.006	10.8 - 397.7	11.6 - 399.9	11.9 - 402.9	11.9 - 405.8	11.7 - 407.8	11.2 - 408.1	10.5 - 406.8	9.9 - 403.6	9.6 - 399.7	9.9 - 397.2
0.007	9.2 - 525.4	9.0 - 523.7	9.1 - 521.5	9.2 - 521.9	8.6 - 524.9	7.7 - 527.6	7.7 - 529.5	8.2 - 529.7	9.0 - 528.4	9.6 - 526.7
0.008	8.8 - 540.7	9.0 - 540.8	8.4 - 541.0	8.1 - 541.2	7.6 - 541.4	7.5 - 541.5	7.9 - 541.6	8.4 - 541.8	8.7 - 541.8	8.7 - 541.8
0.009	12.1 - 614.3	11.9 - 614.5	11.8 - 614.7	11.8 - 614.9	11.5 - 615.1	11.2 - 615.2	11.3 - 615.4	11.4 - 615.5	11.5 - 615.6	11.6 - 615.6
0.01	15.3 - 680.8	15.5 - 680.9	15.4 - 681.2	15.2 - 681.5	15.0 - 681.7	14.7 - 681.9	14.4 - 682.1	14.4 - 682.3	14.7 - 682.4	15.1 - 682.4
0.012	18.4 - 762.2	18.1 - 762.6	18.1 - 763.0	18.3 - 763.4	18.6 - 763.8	19.0 - 764.2	19.3 - 764.5	19.3 - 764.8	18.9 - 765.1	18.4 - 765.3
0.014	34.8 - 781.7	34.9 - 782.2	35.1 - 782.6	34.8 - 783.1	34.4 - 783.6	34.1 - 783.9	33.9 - 784.4	33.7 - 784.8	33.7 - 785.1	33.6 - 785.3
0.016	29.7 - 815.7	29.8 - 816.3	29.8 - 816.8	29.9 - 817.3	29.8 - 817.8	29.8 - 818.3	29.6 - 818.8	29.6 - 819.2	29.5 - 819.6	29.4 - 819.9
0.018	50.3 - 678.4	50.6 - 679.2	51.1 - 679.7	51.6 - 680.1	52.1 - 680.5	52.7 - 680.9	53.2 - 681.4	52.9 - 682.1	52.2 - 682.9	51.6 - 683.6
0.02	37.1 - 278.3	36.9 - 278.3	36.7 - 279.1	36.9 - 279.8	36.8 - 279.4	37.3 - 278.6	36.8 - 278.8	37.2 - 277.9	36.9 - 276.6	36.9 - 278.2

As can be seen in Table 1, min and max values of KLS are changing from plane to plane as well as for different rotations. The first column shows the distance of the planes from bottom of the rotor. Plane 0.001 is the first plane from bottom and 0.02 is the last one.

Eddy size distributions obtained from the simulations are plotted based on Kolmogorov length scale. Calculations were performed for the 10 angular displacements of each plane for 2500 rpm (Figure 2).

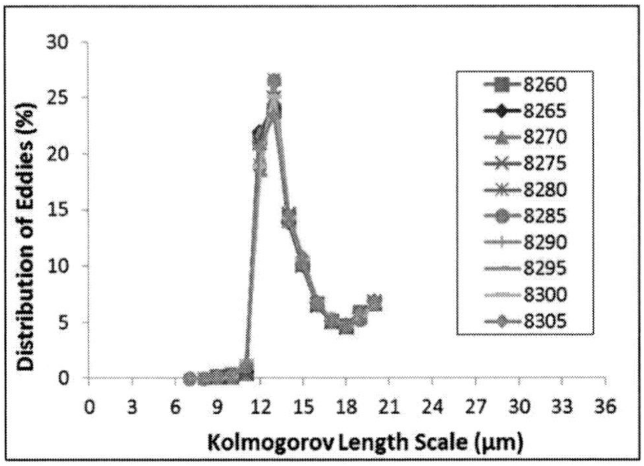

Figure 2.: Distribution of KLS in FDA pump for 2500 rpm for a quarter rotation of pump at different blade configurations which changes from 8260 to 8305.

Size distributions shifted to smaller values with simulations for increasing rotation rate [Figure 2(A)] with minimum and peak values going from about 9 and 13 μm for 2500 rpm. The increased presence of smaller eddies occurs in conjunction with greater hemolysis which can be seen in hemolysis calculations in next section.

4.2. Hemolysis Calculation

Hemolysis calculations are performed for 2500 rpm rotation rate. Hemolysis is calculated using Equation 1 for 10 different blade positions. As can be seen from Table 1 hemolysis values are very low.

Table 2. KLS values on planes for 2500 rpm rotor speed.

Blade Positions	**%Hemolysis (H) values using Eq. 1**
8260	0.5587
8265	0.6115
8270	0.5918
8275	0.5514
8280	0.5149
8285	0.4938
8290	0.4165
8295	0.6596
8300	0.5000
8305	0.5181
Summed H (%)	**5.4164**
H(%) for whole rotation	**21.6656627**
Diluted H(%)	**1.76E-04**

Planes are also created in different distances to examine the KLS distributions and hemolysis values on different plane positions for 2500 rpm. In previous results, planes are created in 0.001 m distances. The distance between planes are increased to 0.0015 m and decreased to 0.0005 m. Moreover, planes are also created on grid points. The hemolysis is quite small for all different cases.

However, the effects of mechanical trauma are completely absent. The cells may be damaged in other ways.

CONCLUSIONS

A validated computational model for hemolysis in turbulent flow has been applied to the FDA critical initiative pump under their specified conditions of operation. The model has been used to predict hemolysis through the pump based on idealized dissipative eddies and their surface area. Results showed that hemolysis values are very small for 2500 rpm. Predictions are to be compared to the experimental findings on release by the FDA.

REFERENCES

1. Givertz MM. Ventricular assist devices important information for patients and families. Circulation. 2011;124:e305-11.
2. Ozturk M, O'Rear EA, Papavassiliou DV. Hemolysis related to turbulent eddy size distributions using comparisons of experiments to computations. Artif Organs. 2015;39:E227-E39.
3. Ozturk M, O'Rear EA, Papavassiliou DV. Reynolds Stresses and Hemolysis in Turbulent Flow Examined by Threshold Analysis. Fluid Mechanics and Heat Transfer in Biological Systems and Clinical/Medical Applications. 2016;1.
4. Ozturk M, Papavassiliou DV, O'Rear EA. An approach for assessing turbulent flow damage to blood in medical devices. Journal of Biomechanical Engineering. 2016;139:011008--8.
5. Burgreen GW, Antaki JF, Wu ZJ, Holmes AJ. Computational fluid dynamics as a development tool for rotary blood pumps. Artif Organs. 2001;25:336-40.
6. Fraser KH, Zhang T, Taskin ME, Griffith BP, Wu ZJ. A quantitative comparison of mechanical blood damage parameters in Rotary Ventricular Assist Devices: shear stress, exposure time, and hemolysis index. Journal of Biomechanical Engineering. 2012;134:081002.
7. Izraelev V, Weiss WJ, Fritz B, Newswanger E, Paterson G, Snyder A, et al. A passively suspended Tesla pump left ventricular assist device. American Society for Artificial Internal Organs. 2009;55:556-61.
8. Morsi YS, Yang W, Witt PJ, Ahmed AM, Umezu M. Numerical analysis of the flow characteristics of rotary blood pump. Journal of Artificial Organs. 2001;4:54-60.
9. Nguyen VT, Kuan YH, Chen PY, Ge L, Sotiropolous F, Yoganathan AP, et al. Experimentally validated hemodynamics simulations of mechanical heart valves in three dimensions. Cardiovascular Engineering and Technology. 2012;3:88-100.
10. Wu J, Paden BE, Borovetz HS, Antaki JF. Computational fluid dynamics analysis of blade tip clearances on hemodynamic performance and blood damage in a centrifugal ventricular assist device. Artif Organs. 2010;34:402-11.
11. B. S, Chua LP, Lim TM, T. Z. Evaluation of the Impeller Shroud Performance of an AxialFlow Ventricular Assist Device Using ComputationalFluid Dynamics. Artif Organs. 2010;60:745-59.
12. Behbahani M, Behr M, Hormes M, Steinseifer U, Arora D, Coronado O, et al. A review of computational fluid dynamics analysis of blood pumps. Eur J Appl Math. 2009;20:363-97.

13. Chiu W-C, M.J. S, Bluestein D. Thrombus Formation Patterns in the HeartMate II VAD- Clinical Observations Can Be Predicted by Numerical Simulations. ASAIO J. 2014;60:237-40.
14. Thamsen B., Blümel B., Schaller J., Paschereit C. O., Affeld K, Goubergrits L, et al. Numerical Analysis of Blood Damage Potential of the HeartMate II and HeartWare HVAD Rotary Blood Pumps. Artif Organs. 2015;39:651-9.
15. US Food and Drug Administration. Computational fluid dynamics: an FDA Critical Path Initiative project. 2013.
16. Stewart SFC, Paterson EG, Burgreen GW, Hariharan P, Giarra M, Reddy V, et al. Assessment of CFD performance in simulations of an idealized medical device: results of FDA's first computational interlaboraroty study. Cardiovascular Engineering and Technology. 2012;3:139-60.
17. Malinauskas RA, Saha A, Sheldon MI. Working with the Food and Drug Administration's Center for Devices to advance regulatory science and medical device innovation. Artif Organs. 2015;39:293-9.
18. Heck ML, Yen A, Snyder TA, O'Rear EA, Papavassiliou DV. Flow-field simulations and hemolysis estimates for the Food and Drug Administration critical path initiative centrifugal blood pump. Artif Organs. 2017;41:E129–E40.
19. Bluestein D. Research approaches for studying flow-induced thromboembolic complications in blood recirculating devices. Expert Review of Medical Devices. 2004;1:65-80.
20. Goubergrits L, Osman J, Mevert R, Kertzscher U, Pöthkow K, Hege H-C. Turbulence in blood damage modeling. Int J Artif Organs. 2016;39:160-5.

575860 Analysis of Pneumonia Associated Volatile Organic Compounds from Bacteria Culture Using Synthesized TiO2 Nanotube Array Sensor and Gas Chromatography/Mass Spectrometry

A vast variety of volatile organic compounds (VOCs) are present in human breath[1]. Microorganisms causing diseases like pneumonia release unique VOCs that are not present in healthy human breath, or they decrease or increase the concentration depending on the pathological state of the patient. Each year about one million people are affected by pneumonia in the Unites States[2]. Early screening for pneumonia would help prevent severe sickness and potential hospitalization later as it would allow healthcare professionals to administer treatment quickly. Monitoring the presence of pneumonia VOCs in the patient's breath is potentially a fast approach for rapid pneumonia screening or diagnosis.

Various VOCs have been reported in literature to be present in the breath of pneumonia patients[3] such as heptane, ethanol, and acetone. Pneumonia is caused by a variety of different bacteria consortium which result in different VOC profiles. Two bacteria known to be associated with pneumonia are *Streptococcus pneumonia* and *Haemophilus influenzae*. In this work we cultured these microorganisms and analyzed the headspace to confirm and establish the VOC profile using gas-chromatography and mass-spectrometry (GCMS).

Titanium dioxide (TIO2) nanotube arrays have been synthesized to detect the mixture of heptane and ethanol in liquid and gas phase with the purpose of utilizing the nanotube arrays as a low-cost screening tool for pneumonia. Our group's previous works show that cobalt functionalized titanium dioxide nanotubes sensors at specific biased voltages detect the presence of tuberculosis VOCs[4-5]. Results from testing non-functionalized titanium dioxide nanotubes and nickel functionalized nanotubes to detect the presence of heptane and ethanol will be presented. Figure.1 shows the preliminary amperometric results of testing a non-functionalized TiO2 sensor in the presence of heptane at 1.5 V. The sensor was exposed to nitrogen gas throw liquid heptane and plain nitrogen gas for a negative check. At a positive biased voltage of 1.5 V, the sensor current-change during heptane exposure was positive. While the current-change was negative when plain nitrogen gas was in contact with the sensor. The experiment is conducted three times.

In this work, GCMS analysis from the VOCs produced by pneumonia cultured bacteria and TiO2 sensors response to pneumonia mimic breath samples will be presented.

References:

Jill D. Fenske & Suzanne E. Paulson (1999) Human Breath Emissions of VOCs, Journal of the Air & Waste Management Association, 49:5, 594-598, DOI: 10.1080/10473289.1999.10463831

Center of Disease Control and Prevention

PoulineMvan Oort, Pedro Povoa, Ronny Schnabel, et al. The potential role of exhaled breath analysis in the diagnostic process of pneumonia—a systematic review. J. Breath Res. 12 (2018) 024001

Bhattacharyya D, Smith YR, Misra M, Mohanty SK. Electrochemical detection of methyl nicotinate biomarker using functionalized anodized titania nanotube arrays. Materials Research Express 2015;2(2):025002.

Bhattacharyya D, Smith YR, Mohanty SK, Misra M. Titania Nanotube Array Sensor for Electrochemical Detection of Four Predominate Tuberculosis Volatile Biomarkers. Journal of The Electrochemical Society 2016;163(6):B206-B214.

Analysis of Pneumonia Associated Volatile Organic Compounds from Bacteria Culture Using Synthesized TiO2 Nanotube Array Sensor and Gas Chromatography/Mass Spectrometry

Yalda Saffary[1], Lani McKinnon[1], Christina Willis[1], Krista Carlson[2], Swomitra Mohanty[1, 2]
[1]Department of Chemical Engineering, University of Utah
[2]Department of Metallurgical Engineering, University of Utah

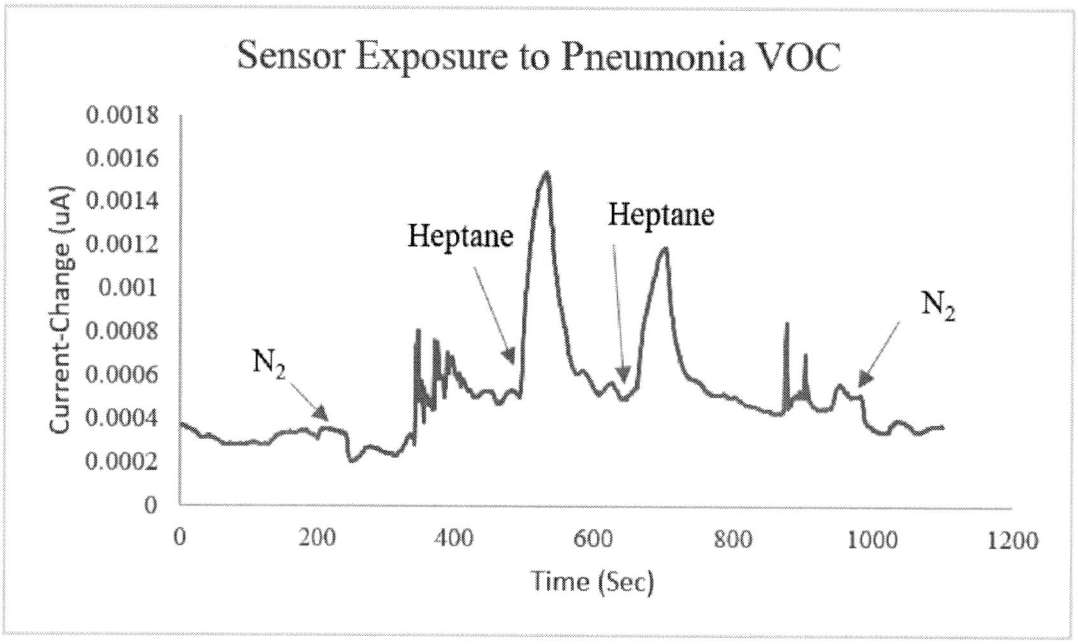

Fig1. An example of current response when a sensor is exposed to nitrogen and heptane

564173 DNA Recovery from Enteric Bacteria for Rapid Diagnostic Genotyping

There is a great need to rapidly identify the organisms in blood infections, and assess their resistance to certain antibiotics, so the correct antibiotic treatment can be implemented from the beginning of therapy. We are developing a microfluidic device that can reduce this diagnostic time from more than 24 hours to under an hour. As part of this process, we need to recover the DNA from bacteria that are captured from infected blood. Bacteria that often contain carbapenam-resistance genes are enteric bacteria, such as *E. coli, Klebsiella pneumoniae*, and *Enterobacter cloacae*. While *E. coli* is relatively easy to chemically lyse in a microfluidic device, *E. cloacae* is fairly difficult. Because the clinician cannot determine *a priori* which bacteria will be present, we have developed chemical lysis procedures that work on all 3 of these bacteria, and potentially on even more species.

Following lysis, the bacteria DNA, both genomic and plasmidic, is collected on silica-coated magnetic beads with a recovery efficiency that depends on the type of bead and the adsorption and elution buffers. With the right chemistry, we have been able to recover about 80% of the DNA. Our presentation describes procedures and chemistry for optimal DNA recover from these medically significant bacteria.

579959 Anti-Amyloid Failures Increase As Aducanumab for Alzheimer's Disease Flops: A Chemical Engineer's Perspective

Although the amyloid (abeta peptide, Aβ) hypothesis is 25 years old, is the dominant model of Alzheimer's Disease (AD) pathogenesis, and currently guides the development of potential treatments, it is still controversial. One possible reason is a lack of a clear mechanistic path from the cleavage products of the amyloid precursor protein (APP) such as soluble Aβ monomer and a series of recently discovered soluble molecular fragments to the deleterious effects on synaptic form and function[1, 2]. Another reason is the successive failure of antibodies against Aβ and secretase inhibitors, highlighted by the recent decision by Biogen and Eisai to terminate two phase III trials of their anti-Aβ antibody, aducanumab[3]. Both biophysical properties of these molecular entities and the balance between production and clearance are considered critical for AD pathogenesis. From a review of the recent literature including aggregation kinetics[4] and structural morphology[4], Aβ clearance[4], molecular simulations[5], long term potentiation measurements with inhibition binding, and the binding of a commercial monoclonal antibody, aducanumab, we hypothesize that the N-terminal domains of neurotoxic Aβ oligomers are implicated in causing the disease. We call this the "N-Terminal Hypothesis for AD"[6]. Recently reported experimental structures also show a flexible/exposed N-terminal region ($A\beta_{1\text{-}14}$) in the disease relevant $A\beta_{1\text{-}42}$ fibril[7]. Additionally, since monomers are not neurotoxic while dimers, trimers etc. are, this suggests that at least two N-termini are necessary for LTP deficit induction. It then follows that the mechanism of toxicity could be due to multivalent interactions between the N-termini (≥2) of Aβ oligomers with glutamate N-methyl D-aspartate receptors. MD simulations demonstrate increased flexibility of the protective versus the causative variants for N-termini in dimers[8]. Taken together, these collective findings strongly suggest that the N-terminus of Aβ oligomers could be causative for AD. Many of the failed antibodies bind to the N-terminus but may not bind multivalently. Since about one in a 1000 antibody molecules pass the blood brain barrier is it possible that too few pass through to exert a long-term affect. Finally, it maybe that AD, which is thought to take ~20 years to mature, is being treated too late and has progressed to far to be reversed. This suggests that early detection is critical[9].

1. Willem, M.; Tahirovic, S.; Busche, M. A.; Ovsepian, S. V.; Chafai, M.; Kootar, S.; Hornburg, D.; Evans, L. D.; Moore, S.; Daria, A.; Hampel, H.; Muller, V.; Giudici, C.; Nuscher, B.; Wenninger-Weinzierl, A.; Kremmer, E.; Heneka, M. T.; Thal, D. R.; Giedraitis, V.; Lannfelt, L.; Muller, U.; Livesey, F. J.; Meissner, F.; Herms, J.; Konnerth, A.; Marie, H.; Haass, C., h-Secretase processing of APP inhibits neuronal activity in the hippocampus. *Nature* **2015,** *526*, 443-447.
2. Welzel, A. T.; Maggio, J. E.; Shankar, G. M.; Walker, D. E.; Ostaszewski, B. L.; Li, S.; Klyubin, I.; Rowan, M. J.; Seubert, P.; Walsh, D. M.; Selkoe, D. J., Secreted amyloid beta-proteins in a cell culture model include N-terminally extended peptides that impair synaptic plasticity. *Biochemistry* **2014,** *53* (24), 3908-21.
3. Mullard, A., Anti-amyloid fauilures stackup as Alzheimers antibody flops. *Nat Rev Drug Discov* **2019,** *18*, 327.
4. Murray, B.; Sorci, M.; Rosenthal, J.; Lippens, J.; Isaacson, D.; Das, P.; Fabris, D.; Li, S. M.; Belfort, G., A2T and A2V A peptides exhibit different aggregation kinetics, primary nucleation, morphology, structure, and LTP inhibition. *Proteins-Structure Function and Bioinformatics* **2016,** *84* (4), 488-500.
5. Das, P.; Murray, B.; Belfort, G., Alzheimer's Protective A2T Mutation Changes the Conformational Landscape of the A beta(1-42) Monomer Differently Than Does the A2V Mutation. *Biophysical journal* **2015,** *108* (3), 738-747.
6. Murray, B.; Sharma, B.; Belfort, G., N-Terminal Hypothesis for Alzheimer's Disease. *Acs Chemical Neuroscience* **2017,** *8* (3), 432-434.
7. Walti, M. A.; Ravotti, F.; Arai, H.; Glabe, C. G.; Wall, J. S.; Bockmann, A.; Guntert, P.; Meier, B. H.; Riek, R., Atomic-resolution structure of a disease-relevant Abeta(1-42) amyloid fibril. *Proc Natl Acad Sci U S A* **2016.**

8. Sharma, B.; Ranganathan, S. V.; Belfort, G., Weak N-terminal interactions for the protective over the causative Aβ peptide dimer mutants. *ACS Chem Neurosci* **2018,** *9* (6), 1247-1253.
9. Hwang, S. S.; Chan, H.; Sorci, M.; Van Deventer, J.; Wittrup, D.; Belfort, G.; Walt, D., Detection of amyloid beta oligomers toward early diagnosis of Alzheimer's disease. *Anal Biochem* **2019,** *566*, 40-45.

575966 (All) Things Merge into One: Viral/Nonviral Chimeric Gene Carriers and Chemically-Induced Extracellular Blebs (EBs) for Efficient, Safe, and Versatile Therapies

An emerging therapy for highly challenging diseases such as cancers and degenerative neurological diseases requires a capability of effectively, specifically, and precisely manipulating the highly complex processes underlying in disease development and progress. Multi-modal therapy overcomes the limitations of narrowly-defined single-modal therapies and is a promising approach to synergistic, targeted, and safe therapeutic outcomes. However, this strategy requires novel platforms encompassing the highly complex demands in compressive functions. Most gene therapies tackle a single pathway, resulting in limited efficiency and making it an ancillary treatment option or approved for rare indications. In this study, adeno-associated virus (AAVs), on which recently approved gene therapy drugs are based, were engineered to encode an apoptotic BIM and further shelled with a MCL-1 siRNA-encapsulating, acid-degradable polymeric layer for avoided immunogenicity during circulation, simultaneous BIM expression and MCL-1 silencing in the target cell, and facilitated eradication of leukemia cells in vitro and in vivo. The same approach is also turned out to be effective against a broad range of cancers such as myeloproliferative neoplasm and non-small cell lung cancer. The extremely high heterogeneity both in physical structure and biological function extracellular vesicles (EVs) and substantially-limited scalability for production have been pivotal challenges in the technology. Induction of EVs by chemical and physical triggers allows the efficient and large-scale preparation of highly homogeneous extracellular blebs (EBs). Via sulfhydryl-blocking, EBs in varying sizes were produced and tested for the delivery of cancer chemotherapeutics[1] and vaccines. Doxorubicin-encapsulating, nano-sized extracellular blebs (DOX-*n*EBs) were prepared by an order of magnitude improved production and they were able to eradicate tumors in vivo with minimized off target effects on non-tumorous tissues. Micro-size extracellular blebs carrying a model SIINFEKL antigen (SIINFEKL-μEBs) were also prepared and they generated effective immunity against the tumor that presents the antigen peptide on the surface. This presentation introduces the current status and changes in EV technology, and how a new approach addresses them in putting it a step forward toward potential clinical translation.

580021 Modulation of Drug and Microbe Transport across the Intestinal Mucosal Barrier

Transport phenomena in the intestinal lumen are highly significant to effective oral drug delivery, nutrient absorption, and interactions between microbes and intestinal tissue. Our laboratory is studying the impact of ingested materials, such as lipids in drug delivery systems or food, on transport and overall absorption of orally delivered compounds. Our approach includes mechanistic studies paired with systems-based models of processes key to overall oral absorption, including dissolution and permeation through intestinal membranes. Quantitative mechanistic understanding of transport phenomena in the drug delivery environment is critical to rational design of drug delivery systems enabling oral delivery of agents that may otherwise require injection. One aspect of permeation through the intestinal membrane, transport through mucus, is being investigated in detail. Results to date highlight the dependence of mucosal permeation on physical and chemical properties of penetrating material (drug, particle, microbe) as well as the significance of intestinal lumen contents in altering the mucus barrier. Our findings suggest that mild stimuli, such as those presented by food, can modulate the intestinal barrier, for example to impact oral drug delivery or microbial invasion.

562766 Customized Medicine for in-Vitro Fertilization (IVF) Using Modeling and Optimal Control

In-vitro fertilization (IVF) is the most common technique in assisted reproductive technology and in most cases the last resort for infertility treatment. It has four basic stages: superovulation, egg retrieval, fertilization, and embryo transfer. Superovulation is a drug-induced method to enable multiple ovulation per menstrual cycle. The success of IVF majorly depends upon successful superovulation, defined by the number and uniformly high quality of eggs retrieved in a cycle. Currently, this step is executed using almost daily monitoring of the follicular development using ultrasound and blood test. The daily dosage of hormones is customized for each patient based on these tests. Although there are the general guidelines for dosage, the dose is not optimized for each patient, and complications, such as overstimulation, can occur. The cost of testing and drugs make this stage very expensive. To overcome the shortcomings of these general guidelines system, a mathematical model and software is developed which can provide a customized model of this stage regarding the size distribution of eggs (follicles/ oocytes) obtained per cycle as a function of the chemical interactions of the drugs used and the conditions imposed on the patient during the cycle, which provide a basis for predicting the possible outcome. The model is based on analogy between IVF superovulation and the particle processes in chemical engineering like seeded crystallization. Customized optimal drug dosage procedures are developed for each patient using optimal control methods with the targeted follicle size distribution (FSD) as the expected outcome. This paper describes the theory, model, and the optimal control procedure for improving outcomes of IVF treatment for one of the two protocols used in real practice. The validation of the procedure is performed using clinical data from the patients' previously undergone IVF cycles. Customized patient-specific model parameters are obtained by using initial two-day data for each patient. Subsequently, this model is used to predict the FSD for the remaining days of the cycle. This procedure was conducted for 49 patients. The results of the customized models are found to be closely matching with the observed FSD.

These results thus validate the modeling approach and consequently its use for predicting the customized optimal drug dosage for each patient. Using the customized model and the optimized dosage, the FSD at the end of the cycle was determined. A small double-blind clinical trial was also conducted in India. The results from the trial show that the dosage predicted by using the model is 40 % less than the suggestion made by the IVF doctors. Additionally, the number of mature follicles obtained at the end of the cycle using the model-predicted dosage was significantly higher than the physician suggested dosage. These results were consistent with all patients in this clinical trial. The testing and monitoring requirements for patients using optimized drug dosage is reduced by 72%. We are currently working on patients from United States for both protocols and the results are promising.

574642 Computational Fluid Dynamics Simulations of the Human Carotid Artery to Predict Strokes in Patients with Carotid Artery Disease

Computational Fluid Dynamics Simulations of the Human Carotid Artery to Predict Strokes in Patients with Carotid Artery Disease

Anna K. Weldy, Amanda M. Forti and David G. Foster

Department of Chemical Engineering

University of Rochester

Rochester, New York

The purpose of this study is to investigate the potential for Computational Fluid Dynamics (CFD) simulations to predict stroke in patients with carotid artery disease. CFD was used to study the flow patterns using patient's carotid artery scans paying specific attention to the geometry, velocity profiles, streamlines and pressure gradients. Using these data, we were able to computationally simulate individual patient's unique blood flow patterns and pressure gradients based on computerized tomography angiography (CTA) for geometry, and ultrasound data for velocity and pressure data. First, we obtained patient results coupled with patient outcomes for stroke, to develop the understanding of which flow patterns and geometries were most likely predictors of adverse outcomes. Next, we obtained additional patient data without the patient outcomes, and compared our predicted outcome to the actual outcome. These correlations are then used to modify our data analysis to ensure a more accurate prediction of patient outcomes based on the CFD simulations. This paper will overview our work and the present status toward an accurate prediction of stroke in patients using CFD simulations.

576466 Personalized Treatment of Sickle Cell Disease Using Hydroxyurea Pharmacokinetic-Pharmacodynamic Modeling

Sickle Cell Disease (SCD) is the most common inherited blood disorder, affecting approximately 100,000 people in the United States. Hydroxyurea (HU) is the only FDA approved disease-modifying drug for SCD. The current treatment method involves daily oral dosing of hydroxyurea over a patient's lifetime. There are many challenges associated with HU treatment. Firstly, due to different genetic and phenotypic makeups, there are wide interpatient variability in pharmacokinetic-pharmacodynamic (PK-PD) profiles which leads to variability in maximum tolerated dose (MTD). Therefore, HU dosing and escalation requires frequent visits to the hospital and usually takes 6-9 months to reach the MTD. Secondly, there is a need for a better biomarker to indicate treatment efficacy. The biomarkers currently in use are fetal hemoglobin (HbF) and mean cell volume (MCV) of red blood cells, both of which increase with HU treatment. But it is found that they take about 3 months or longer to reach a steady-state value, thus implying a delay in the clinical decision making of whether the dose should be increased or not. Lastly, non-adherence presents a major challenge. It is very difficult to differentiate between treatment inefficacy and non-adherence, which can result in underdosing, or overdosing. Due to these challenges, there is a lack of optimal treatment for individual patients. A mathematical model will be useful in addressing all the foregoing challenges through timely and personalized prediction of optimal dose, which is addressed in this work via a PK-PD model.

The PK model consists of three compartments: gastrointestinal tract, plasma, and liver where the drug gets metabolized to nitric oxide and its derivatives. This model describes the temporal distribution of HU and its metabolite in the compartments. The PD model is described by HU toxicity and efficacy models. To capture HU toxicity as a function of blood cells count, a population balance model is formulated with cell age and volume as the internal coordinates. This model also predicts MCV of red blood cells and reticulocytes which is used to calculate treatment efficacy. To quantify HU efficacy as a function of HbF, a signaling model is developed that describes HU-induced HbF activation through the NO-cGMP pathway. By integrating these models, an individual patient's response (PK-PD variable dynamics) is predicted, which is used to compute the optimal personalized dose. This dose is calculated by maximizing efficacy (HbF, MCV level) and minimizing toxicity (blood cells count). This model is also being used to predict patient adherence. Using a probabilistic distribution for non-adherence, the effect of missing the dose on PK-PD trajectory is analyzed and the expected optimal dose is computed for non-adherent patients. In conclusion, an integrated PK-PD model is developed which can predict optimal HU dose by capturing the variant responses of SCD patients.

566711 Antioxidant-Encapsulating Nanoparticles for the Treatment of Glutamate Excitotoxicity

Introduction: Despite the neurological advancements of the past century, neurological diseases still have no effective cures, and cost nearly $800 billion annually in the United States alone. Most neurological diseases, including stroke, traumatic brain injury, and Alzheimer's, exhibit glutamate excitotoxicity that perpetuates injury. Glutamate is a major neurotransmitter used for neuronal communication but leads to neuronal death when in excess by over-activating glutamate receptors. The mechanism behind excitotoxicity involves an excessive influx of calcium and generation of reactive oxygen species (ROS), and subsequently leads to neuroinflammation in many cases. After death, neurons release their glutamate depots, promoting further neighboring neuronal death. Stopping the excitotoxic cascade is crucial for preventing additional damage and promoting recovery. However, the many obstacles to drug delivery to the brain, including systemic clearance, blood-brain barrier (BBB) transport, and brain parenchyma penetration, as well as the complex etiology of disease initiation and propagation complicate the study and effective treatment of excitotoxicity. The use of polymeric nanoparticles for encapsulating and delivering therapeutic cargo can help overcome several of these barriers. Still, discerning whether a therapeutic failed due to ineffective biochemistry as opposed to poor delivery is challenging in the complex *in vivo* environment. Testing therapeutics on simplified, yet representative models, provides a systematic approach towards evaluating drug efficacy. By utilizing *ex vivo* whole hemisphere brain slice models, we can assess therapeutic efficacy, nanoparticle biocompatibility, as well as probe for deeper understanding of excitotoxicity mechanisms in high-throughput fashion.

Materials and Methods: We developed an organotypic whole hemisphere brain slice model of excitotoxicity. In this model, we incubated slices in excitatory conditions and evaluated cytotoxicity via lactate dehydrogenase (LDH) release, mRNA profiles with real-time polymerase chain reaction (RT-PCR), and microglial and mitochondrial morphologies with immunohistochemistry. We next determined the therapeutic efficacy of antioxidant enzymes, namely superoxide dismutase (SOD) and catalase. Additionally, we formulated antioxidant-encapsulating poly(lactic-*co*-glycolic acid)-*block*-poly(ethylene glycol) (PLGA-PEG) polymeric nanoparticles (AE-NPs) using the double emulsion method, and evaluated their therapeutic potential and enzymatic activity in degradative conditions.

Results and Conclusions: We obtained a characteristic profile of cell viability and disease environment for excitotoxicity-induced brain slices. After application of SOD, we observed decreased cell death and a return to healthy mitochondrial and microglial morphologies. This effect was not seen with catalase treatment. Additionally, we formulated nanoparticles with an antioxidant/PLGA core and a dense PEG coating to achieve a hydrophilic, bio-inert surface. The AE-NPs exhibited a near-neutral surface charge and sub-100 nm size, and extended enzymatic activity compared to un-encapsulated enzyme *in vivo* and in the presence of pronase *in vitro*. We further used *ex vivo* slices to identify PEG toxicity of our nanoparticles, and for optimization of a biocompatible formulation. Our work provides evidence of SOD efficacy and supports the use of NPs to better deliver therapeutics to combat excitotoxicity. Beyond promoting SOD-NPs against excitotoxicity, our findings encourage the systematic approach of screening therapeutics on *ex vivo* brain slice models to isolate, understand, and overcome neurological disease processes. The slice model retains the multi-cell type complexity of the brain while bypassing systemic and BBB obstacles. By evaluating efficacy and biocompatibility without confounding delivery issues, we can focus on individual therapeutic development bottlenecks to ultimately design an effective therapeutic against glutamate excitotoxicity.

576602 Engineering Variable Lymphocyte Receptors to Target Blood Brain Barrier Disruption Via Exposed Neural ECM

Introduction: Normally the blood-brain barrier (BBB) prevents interaction between blood components and brain tissue, including neural extracellular matrix (ECM). However, certain neurological diseases, such as traumatic brain injury, stroke, and brain tumors, exhibit pathologic BBB disruption that is localized to the disease region, thus, exposing normally inaccessible neural ECM to circulating blood components. We propose exploiting differential exposure of neural ECM as a method for targeting therapeutics to pathologies that induce BBB disruption. Targeting therapeutics to the site of BBB disruption, as opposed to utilizing differential expression of cellular receptors, could have multiple benefits including targeting both diseased and supporting stromal cells, identifying multiple sites of BBB disruption within a single patient, and have generalizable application to multiple types of neuropathologies that exhibit a disrupted BBB phenotype.

Methods: We developed a semi-high throughput screen to identify Variable Lymphocyte Receptors (VLRs, the antigen recognition system from Lamprey) that demonstrate high specificity for neural ECM. Initial VLR clones underwent further protein engineering for selective neural ECM binding using in vitro murine tissue accumulation and immunostaining assays. Results were confirmed in vivo, using manitol disruption and murine glioblastoma models.

Results: Multiple neural ECM-binding VLRs were identified. The lead candidate identified after protein engineering, named P1C10, demonstrates diffuse binding to parenchymal neural ECM, without detectable binding to ECM from clearance organs. Additionally, P1C10 demonstrates nanomolar affinity for in vitro derived neural ECM. P1C10 accumulated in murine brains exposed to mannitol as well as at the tumor site of intracranial GL261 and U87 murine models of GBM. Finally, doxorubicin-loaded liposomes targeted with P1C10 conferred a significant survival benefit to mice bearing intracranial U87 GBM.

Conclusions: Here we present the rationale and proof-of-concept demonstration for targeting sites of BBB disruption via exposure of neural ECM. Novel protein engineering methods were used to screen and modify ECM-binding VLRs. Neural ECM-binding VLRs selectively target intracranial sites of GBM and functionally deliver therapeutic cargo. Thus, this new physiological-based targeting scheme may be well suited to target sites of BBB disruption in a variety of neural pathologies.

Acknowledgements: This work was supported by National Institutes of Health grants NS091851 (E.V.S and B.R.H.) and NS099158 (E.V.S.), a Falk Medical Research Trust Catalyst Award (E.V.S. and J.S.K.) and a Defense Threat Reduction Agency grant HDTRA1-15-1-0012 (E.V.S.).

568793 Effect of Screening Compliance on Optimal CRC Screening Policies

Effect of Screening Compliance on Optimal CRC Screening Policies

Session: Chemical Engineers and Public Health

David Young, Selen Cremaschi
Department of Chemical Engineering, Auburn University, Auburn, AL, USA

Abstract

It was estimated[1] that in 2010 costs associated with colorectal cancer (CRC) care totaled around $14 billion for the U.S. alone. CRC is also the third most common and the second most deadly form of cancer worldwide according to the World Health Organization[2]. Deaths associated with CRC are closely linked to how far the cancer has progressed within the individual, with late stage 5-year survival rate for CRC being around 12 %, whereas early stage 5-year survival rate being upwards of 90 %. Therefore, early detection of CRC and its precursors is an effective mitigation technique to reduce the overall impact and the financial burden of CRC on society. Early detection is achieved for CRC through screening an asymptomatic individual using a specified screening test(s) throughout an individual's life. What screening test(s) to perform and at what age(s) to perform these tests are referred to as a screening strategy.

The impact of different screening strategies on total CRC costs and life years gained have been studied for general populations by what-if analysis utilizing CRC microsimulations. Microsimulations are computer models that simulate numerous entities and allow researchers to study the interactions and impacts that a decision or policy has on those entities[3]. In the case of a CRC microsimulation, the entities are individuals, and the decisions are screening strategies to apply to a population of those individuals. Each individual has the progression of CRC described within their lifetime within a section of the simulation known as the natural history portion. This portion models the adenoma-carcinoma sequence allowing for some of the individuals to develop CRC. The screening strategy is implemented within the screening portion of the simulation. This portion incorporates screening compliance for the individuals, and it then applies the desired screening to the individual, if that individual is compliant[4]. Knudsen *et al.*[5], Goede *et al.*[6], and Aronsson *et al.*[7] are recent studies that analyze the impact of screening strategies on CRC costs and life years gained. In these studies, only one implementation of screening compliance model was considered, e.g., flat compliance rate with an individual either always or never compliant. The results of their sensitivity analysis revealed that the benefits of screening are noticeably affected by varying compliance. Furthermore, one compliance model may not be adequate for representing the compliance behavior of all individuals in a population. To the best of our knowledge, there are no studies that systematically compare the impacts of different screening compliance models on the cost-effectiveness of and life years gained due to CRC screening strategies.

In this study, we coupled a derivative free optimization (DFO) technique with a microsimulation model of CRC to study the effects of different screening compliance models on the optimal CRC screening strategy, and cost-effectiveness and life-years gained due to that strategy. Our previous work[8] showed that the coupled DFO-CRC microsimulation framework enables the identification of the optimal CRC screening strategy for cost-effectiveness. Derivative free optimization is a field of optimization in which the optimization algorithm does not know the functional form of the objective function for the problem it is solving, treating it as a black-box model. The algorithm can obtain the value of the objective function for any valid set of decision variable values by directly evaluating the black-box model. The algorithm then uses specific search strategies to determine the next set of decision variable values, iteratively trying new sets to converge to the optimal objective value. In our framework, we have a CRC microsimulation as our black-box model, with the decision variable as the screening strategy to be evaluated. The output, or objective, of our black-box model, is the difference in the total cost associated with screening and treating

CRC for a population that is screened compared to the same population if they never partook in a screening regime. The objective also incorporates the value of the quality of life years gained through screening by putting a monetary value of $100,000 per quality adjusted life year gained through screening, a value used as a threshold for evaluating the effectiveness of various CRC screening strategies[9]. This framework is used to analyze and compare four different compliance models in terms of their effects on the cost-effectiveness and life years gained: 1) perfect compliance, i.e. every person follows the recommended strategy, 2) flat compliance, i.e., an individual is either perfectly compliant or never compliant with the recommended strategy, 3) sometimes, always or never compliant, where the sometimes compliant individuals follow a strict annual compliance, where the ages at which the strategy recommends screening are the only ages the individuals are allowed to be screened, 4) sometimes, always or never compliant, where the sometimes compliant individuals follow a loose annual compliance, where screening can occur at any age between the starting and ending ages, but must be at least the set number of years apart. Within each of these models, the compliance rate is based on the test type of the recommended screening strategy. This talk will give a brief overview of the coupled DFO-CRC microsimulation framework and discuss in detail the changes in the optimal screening strategies and the benefits to the society under different screening compliance models.

References

1. Mariotto, A. B., Robin Yabroff, K., Shao, Y., Feuer, E. J. & Brown, M. L. Projections of the cost of cancer care in the United States: 2010-2020. *J. Natl. Cancer Inst.* **103**, 117–128 (2011).
2. World Heath Organization. Cancer. (2019). Available at: https://www.who.int/news-room/fact-sheets/detail/cancer. (Accessed: 27th February 2019)
3. Orcutt, G. H. A new type of socio-economic system. *Rev. Econ. Stat.* 116–123 (1957).
4. Rutter, C. M. & Savarino, J. E. An evidence-based microsimulation model for colorectal cancer: Validation and application. *Cancer Epidemiol. Biomarkers Prev.* **19**, 1992–2002 (2010).
5. Knudsen, A. B. *et al.* Cost-effectiveness of computed tomographic colonography screening for colorectal cancer in the medicare population. *J. Natl. Cancer Inst.* **102**, 1238–1252 (2010).
6. Goede, S. L. *et al.* Cost-effectiveness of one versus two sample faecal immunochemical testing for colorectal cancer screening. *Gut* **62**, 727–734 (2013).
7. Aronsson, M., Carlsson, P., Levin, L., Hager, J. & Hultcrantz, R. Cost-effectiveness of high-sensitivity faecal immunochemical test and colonoscopy screening for colorectal cancer. *Br. J. Surg.* **104**, 1078–1086 (2017).
8. Young, D. & Cremaschi, S. A Simulation-based Optimization Approach to Develop Personalized Colorectal Cancer Screening Strategies. in *Computer Aided Chemical Engineering* **44**, 2125–2130 (Elsevier, 2018).
9. van Hees, F. *et al.* The Appropriateness of More Intensive Colonoscopy Screening than Recommended in Medicare Beneficiaries: A Modeling Study. *JAMA Intern. Med.* **174**, 1568–1576 (2014).

576380 An Approach Towards Method Development for Untargeted Urinary and Serum Metabolite Profiling in Metabolomics Research, As a Tool in Exposome Studies, Using Uplc-Q-TOF/MS

The exposome, refers to the total of exposures from the time of conception onwards, including the exogenous and endogenous exposures and more of that the different modifiable risk factors that predetermine to and predict disease. In line with exposome, exposome science aims at capturing the mechanistic processes that describe the source-to-dose continuum. Towards this aim, several methodological tools are employed related to environmental and human biomonitoring. Biological perturbations at different levels of biological organization are interpreted with multiple-omics (including genomics, transcriptomics, proteomics, and metabolomics) and post-omics technologies including epigenomics.

Furthermore the identification of the perturbations, results in the establishment of presumable pathways of toxicity. Pathways are been verified by targeted multi-omics and functional assays. Advanced bioinformatics tools, such as support vector machines and clustering algorithms and systems biology models, allow us to identify the functional links among the data derived from high throughput testing platforms and disease phenotypes providing thus phenotypic anchoring of the mechanistic hypotheses made earlier.

Here in this study we are focusing in the urinary and serum metabolomics profiling in metabolomics research, as a tool in exposome studies, using UPLC-Q-TOF/MS.

Global metabolic profiling (metabolomics/metabonomics), using LC–MS as an analytical platform, attracted great deal of interest in recent years in toxicological and pharmaceutical research, as well as in disease biomarker discovery. Although the last years progress has been made, a) in generating more or less comprehensive metabolite profiles, b) in data analysis, and c) in biomarker identification the actual methodology for untargeted profiling using LC–MS is in continuously development. The most difficult and the same time the most interesting in untargeted metabolite profiling is to optimize the different experimental steps, for studying thousands of unknown metabolites in biological samples. Both analytical and biological variability that exists, must be evaluated.

In this study we aimed to evaluate various sample pre-treatment and LC conditions in order to optimise a method for untargeted urinary and serum metabolite profiling that would yield high quality and reproducible data using UPLC/QToF MS. We tested the effects of varied experimental procedures on total ion chromatograms, total number of features and the repeatability of selected endogenous metabolites. We also investigated different approaches to evaluate and monitor the performance of analytical platforms used in generic metabolite profiling as well as testing the stability of biofluids under various sample handling and storage conditions. Major importance was the use of QC samples in metabolic profiling analyses. QC samples are the samples that became of a pool sample of the matrix to be analyzed and prepared by taking small aliquots of each of the study samples and mixing them to form a single “Pool”. The Pool sample and furthermore the QC samples, will contain a mean concentration of all of the components that are present in the samples under investigation. The novel in this method is the fact that internal standards have been used in all QC samples in order to strengthen the use of the QC samples. The QC samples have been placed at regular intervals throughout the analytical run in order to monitor different factors though the analysis. When combined with other measures, such an approach can rapidly be used to determine, if an analytical run has performed as it should, and the data is reliable enough to take forward into more detailed multi variate statistical analysis or, alternatively, that there has been a problem that requires investigation and resolution. In reality, this approach adapts established procedures that are followed in bioanalytical method validation to reach as much as possible the aim of characterizing the results obtained in terms of their reliability. What is next? Data preprocessing and data processing.

The main goals in spectral processing is to correctly arrange the huge amount of raw data generated by the MS files into a 2D matrix, and the improvement of signal quality and the reduction of the possible analytical and biological biases present in the raw data. There are several open source software available tools providing different methodological options for spectral processing. The most important criteria for the selection of the suitable tool are 1) the analysed biological sample, 2) the analytical technology used, and 3) the tool that has been chosen for further data analysis or pathway analysis, since the generated feature matrix should be in a format compatible with the tool. For the LC-MS data preprocessing a combination of R packages, including, xcms and limma, were used.

Regarding the LC-MS data analysis, the .d files generated from negative and positive ionization are treated as two different experiments. The parameters for noise removal, mass detection, deconvolution, data transformation, data reduction etc. were set based on the behavior of QC samples, because QC samples providing an average of all the metabolomes analysed in the study, as mentioned before. The asymmetric baseline corrector was used as the correction method, and for peak detection the centroid was used, since data were already centroided. In order to define the mass error for chromatogram builder, internal standards spectrums, such as caffeine or reserpine, were used as a reference. This is the reason why reference samples in the beginning, in the middle and at the end of analysis. Local minimum algorithm was selected due to low noise level, for the deconvolution. After deisotoping, alignment and gap-filling using the algorithm peak finder, a file containing the m/z values, retention times, and peak area for each detected peak, was exported in CSV format. Beside the careful design of the samples acquisition process, the careful cleaning and maintenance of the equipment before a batch analysis, to obtain consistent variables the resulting matrix was further reduced by the 80% rule. The 80% rule should not be applied to the QC samples rather to the problem samples, to avoid the loss of important metabolites reflecting to the biological status of the system. The instrument and overall process variability were determined by calculating the median RSD for all the endogenous metabolites, in case of the presented cohorts.

The next step in the downstream bioinformatics analysis was pathway mapping that revealed the roles that metabolites play in relation to each other and in biological aberrations, and EWAS analysis to draw the links between in utero exposure to metals, metabolic pathway deregulation, and clinically observed phenotypes of neurodevelopmental disorders. In particular, the logistic regression and FDR was carried out using the 'X-Wide Association Analyses' package, following the EWAS framework.

The application of this method was successfully used in two big cohord studies in the frame of HEALS (Health and Environment-wide associations via Large population Studies) project, called Repro PL and PHIME.

557702 Highly Conductive, Stretchable, and Biocompatible Nanocomposite for Electromechanical Cardioplasty

Intrinsically stretchable conductors form a vital component of advanced soft bioelectronic devices. And novel nanocomposites based on conductive nanomaterials have been used in diverse applications of implantable and wearable bioelectronics. Among many nanomaterials for the composites, silver nanowires (Ag NWs) have been popular. However, achieving highly conductive and soft composites simultaneously is challenging. Because bioelectronics is necessarily exposed to biofluids, preventing Ag NW oxidation and Ag ion leaching is a significant challenge. Here we have achieved a highly conductive, biocompatible, and soft nanocomposite by using silver-gold (Ag-Au) core-sheath NWs and polystyrene-butadiene-styrene (SBS) elastomer. We synthesized ultralong Ag NWs encapsulated with a smooth and uniform Au sheath and then mixed them with the polymer. Phase separation of Ag-Au NWs and SBS occurs, which forms microstructures in the composite, reduces Young's modulus, and increase softness and conductivity of the composite. We used the nanocomposite to fabricate a customized multi-channel soft cardiac mesh for mapping the diseased swine heart epicardially and applying customized feedback stimulation therapy. The global resynchronization therapy based on the multichannel pacing improved the pumping capability of the animal heart dramatically.

576631 Core-Shell Gallium/Polymer Microstructure for Thermal-Responsive Implantable Electrode

Implantable microelectrodes are essential tools for recording and stimulation of excitable tissues that have attracted great interest because of their potential to revolutionize the treatments of many diseases. Mechanical rigidity is essential to assist the device implantation process. However, high stiffness often causes tissue damage and ultimately induces a chronic inflammatory response, which generates gliosis around the implanted devices. Therefore, the ideal implantable electrode needs to be mechanically rigid during the tissue insertion procedure, then adapt to flexible mechanical properties in the physiological condition. Recently, stimuli-responsive materials, such as cellulose whisker and polymer gel, are researched to generate sharp mechanical property transition with water swollen. However, these mechanical transitions are confined solely for the substrate, not for the electrode material that causes delamination between those.

Here, Gallium (Ga) is introduced as an electrode material because it has a unique melting point at 29.36 °C. This indicates Ga will be a rigid solid at room temperature (Young`s modulus of 10 GPa) and a liquid (no mechanical strength) at body temperature. This provides an ideal sharp mechanical transition, rigid to soft, between 23°C to 37°C for implantable microelectrodes. In this study, we developed a novel manufacturing process to create a gallium/polymer core-shell structure. The shell can be used as not only a container to prevent leaking when Ga changed to a liquid state but an insulator. Polyether block amide (PEBAX) was selected as a shell material because of its flexibility and biocompatibility. Lastly, Poly(3,4-ethylene dioxythiophene) (PEDOT) was chosen for the ion detecting site, which helps stimulation property as well as recording. By the help of those characteristics of each material, we fabricated the 50 μm diameter thermal-responsive microelectrodes.

We evaluated the mechanical and electrochemical performance of this core-shell structure for implantable electrodes. In first, the microelectrodes have a stiffness enough to penetrate brain tissues. In addition, Young`s modulus is decreased by three orders of magnitudes from keeping at room temperature to body temperature. The result demonstrates that the microelectrodes have a lower stiffness than brain tissue, which means the devices are helpful to minimize implant history. Lastly, we confirmed that their electrochemical properties such as impedance and capacitance showed suitable performance to apply electrodes, even they have electrochemical stability for four weeks. All abilities including sharp mechanical transition in a temperature change may find uses in chronical neuro-prosthetic devices.

574751 Morphing Electronics for Growing Tissue

Current state-of-the-art implantable electronics does not yet accommodate the physical growth of the biological tissue and organ during development. Here, we address this limitation with a new approach, namely Morphing Electronics (termed MorphE) that can suitably adapt to the nerve tissue growth during adolescent development *in vivo*. In specific, we employed viscoplastic electronic materials to minimize the stress and tissue constraint on rat's growing nerves. Achieved capabilities in water-insensitive and self-healing enabled efficient vertical self-bonding for multilayered MorphE formation, in which it contained viscoplastic electrodes and strain sensors. More importantly, it also allowed for an easy reconfigurable adaptation for individualized implantation procedures. Furthermore, our fabricated soft and form-accommodating MorphE device showed no evident damages on the rat's growing nerves, thus allowing for a stable functional interface for chronic neural stimulation and monitoring without disruption of rat's functional behavior.

575095 Soft and Biostable Silver - Gold Core - Shell Nanowire Nanocomposite Film

Commonly, implantable bioelectronics are fabricated from rigid materials that can lead to inflammation and performance degradation during prolonged use. New approaches must be taken to produce devices with softer mechanical properties more closely matching their tissue environment. By making a nanocomposite material, both electrical and mechanical properties may be manipulated to allow for a more compatible implantable device. Limitations include making a nanocomposite material with high enough of a conductivity while still expressing softer properties. Benefits include the device adapting more suitably soft and stable properties attributable to the polymer matrix. By taking a systematic approach, a range of conductivities and softnesses have been achieved and characterized. Further investigation of the inflammation processes around these devices are required to establish their prospects as long-term neural electrodes.

578765 Nanowire Arrays Restore Vision in Blind Mice

The restoration of light response with complex spatiotemporal features in retinal degenerative diseases towards retinal prosthesis has proven to be a considerable challenge over the past decades. Herein, inspired by the structure and function of photoreceptors in retinas, we develop artificial photoreceptors based on gold nanoparticle-decorated titania nanowire arrays, for restoration of visual responses in the blind mice with degenerated photoreceptors. Green, blue and near UV light responses in the retinal ganglion cells (RGCs) are restored with a spatial resolution better than 100 μm. ON responses in RGCs are blocked by glutamatergic antagonists, suggesting functional preservation of the remaining retinal circuits. Moreover, neurons in the primary visual cortex respond to light after subretinal implant of nanowire arrays. Improvement in pupillary light reflex suggests the behavioral recovery of light sensitivity. Our study will shed light on the development of a new generation of optoelectronic toolkits for subretinal prosthetic devices.

566761 Viscoelastic Aerosol: Novel Inhalation and Oro-Nasal Routes to Drug Delivery

PARC, A Xerox Company is developing a novel approach to aerosol generation from viscoelastic fluids called Filament Extension Atomization (FEA). The method is specifically suited for spraying highly viscoelastic, strain hardening fluids such polymers and biomacromolecules which are often difficult to spray using conventional methods that are viscosity-limited. The FEA approach employs the intrinsic elastocapillary instability in these strain hardening fluids to generate narrowly-dispersed micron-sized droplets of the fluid that can used for various applications. FEA can generate spray from a wide range of fluids with viscosities ranging from 1mPa-s up to 600 Pa-s and higher with tunable droplet sizes down to the 3-5 microns range. In this contribution, we will highlight its potential applications in drug delivery - specifically in unlocking new delivery routes (oro-nasal, transmucosal, inhalation) for bioactives.

577056 Fractionation of Extracellular Vesicles into Molecularly Homogeneous Subpopulations

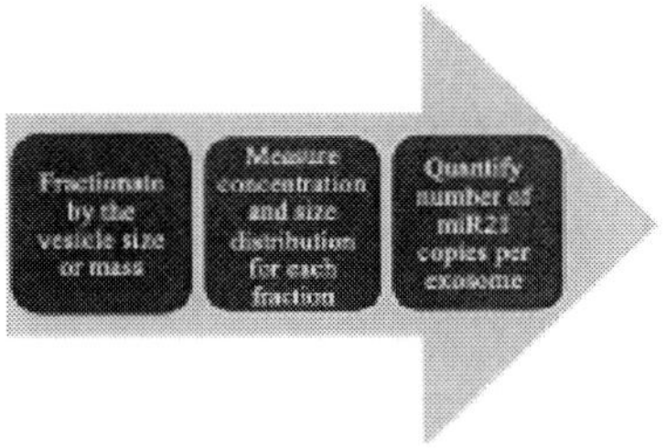

Figure 1: Fractionation and quantification workflow.

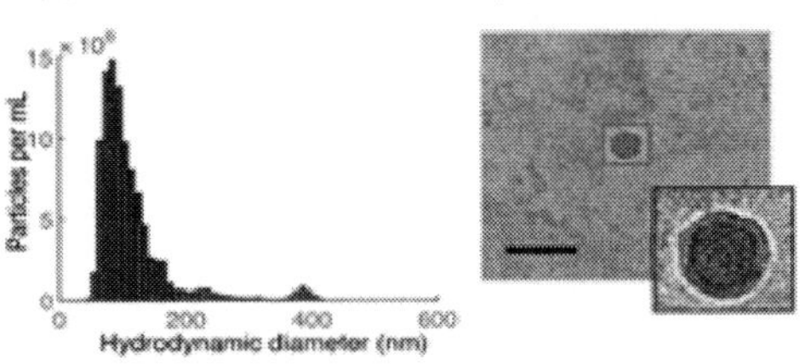

(c)

% Expressed	Biomarker
33	CD63
0	EPCAM
57	ANXA5
57	TSG101
0	Blank
100	Positive

Biomarker	% Expressed
Positive	100
GM130	0
FLOT1	27
ICAM	15
ALIX	12
CD81	45

Figure 1: Quantification of the vesicle size distribution, their morphology, and the expression of the exosomal protein biomarkers.

The heterogeneity of extracellular vesicles (EVs) in biofluids is, both, the basis of liquid biopsy, in which the molecular composition of exosomes secreted by afflicted cells is used in disease detection and characterization, and the fundamental challenge. At an early stage of a disease, exosomes secreted by few afflicted cells form an exceedingly small subpopulation that is difficult to detect and characterize by downstream molecular analysis. At all conditions, the molecular signaling intermediated by heterogeneous exosomes is complex. Molecules transferred by molecularly distinct exosomes may transmit competing signals. For example, within the population of the exosomes isolated from blood of a cancer patient, some may contain microRNAs associated with tumor proliferation, while others may include tumor-suppressing microRNAs. By separating the sample into subpopulations with the reduced molecular heterogeneity, it will become possible to repeatably study signaling interactions and identify biologically active compound associated with specific functional outcomes.

For a specific case of microRNA miR21 cargo in MCF-7 exosomes, we demonstrate a widely different abundance of miR21 in MCF7 exosomes along different biophysical vectors. We show the feasibility of obtaining subpopulations of intact exosomes with higher uniformity in miR21, including the capability to enrich or deplete the miR21 expression from the original population.

EVs were isolated from the growth medium of MCF-7 cells by precipitation. The exosomal biomarkers confirming the enrichment of exosomes in the isolated population of EVs was confirmed by the protein biomarker array, Figure 1(c). The nanoparticle tracking analysis (NTA) was used to verifu that the isolated vesicles are in the range of hydrodynamic sizes expected for the exosomes, Figure 1(a). The bilayer morphology of the vesicles was confirmed by cryo-TEM imaging which shows the expected 3-4 nm thickness of the membrane, Fig. 1(b).

The fractionation of EVs by their sizes and buoyant density was performed by asymmetric and centrifugal field flow fractionations (AFFF and cFFF), respectively. Figure 2 summarizes the fractionation and the quantification workflow used in this study. The absolute quantification of the miR21 copies per vesicle was obtained by digital droplet PCR (ddPCR). The ddPCR results for the miR21 abundance in different factions are shown in Figure 3. The same results expressed in miR21 copy number per vesicle units is shown in Figure 4 for the size fractions and in Figure 5 for mass fractions.

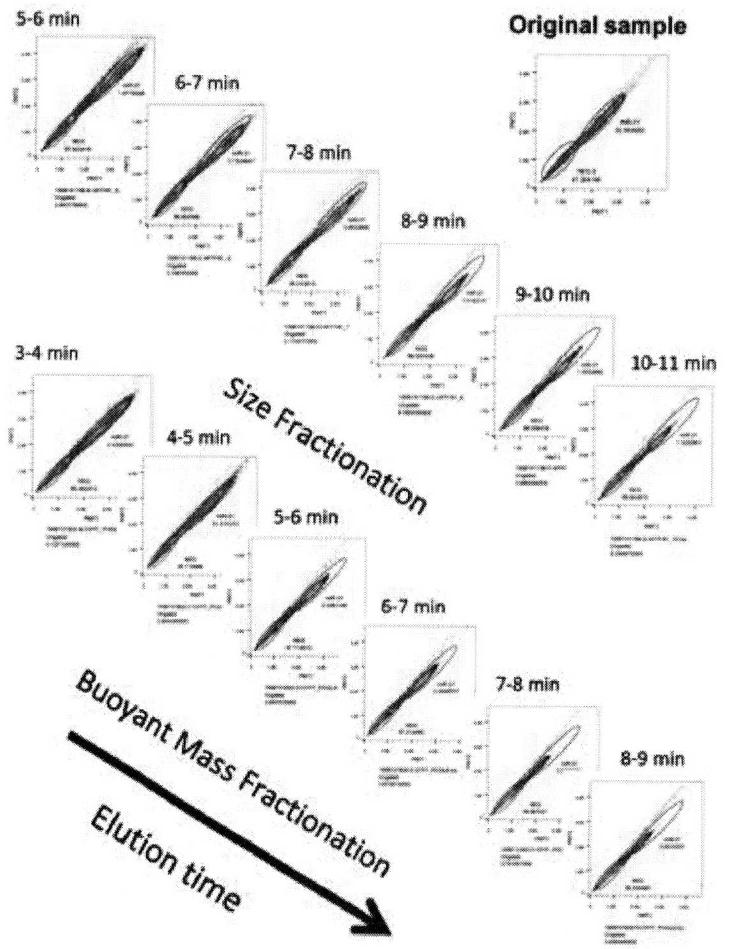

Figure 1: miR21 abundance in different size and mass fractions of MCF-7 exosomes obtained by ddPCR.

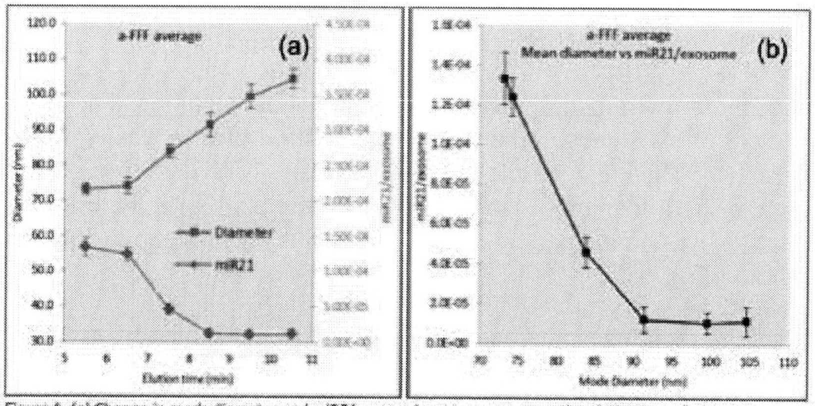

Figure 1: (a) Change in mode diameter and miR21 expression per exosome vs. the elution time during size fractionation by AFFF. (b) miR21 expression per exosome vs. the mode diameter.

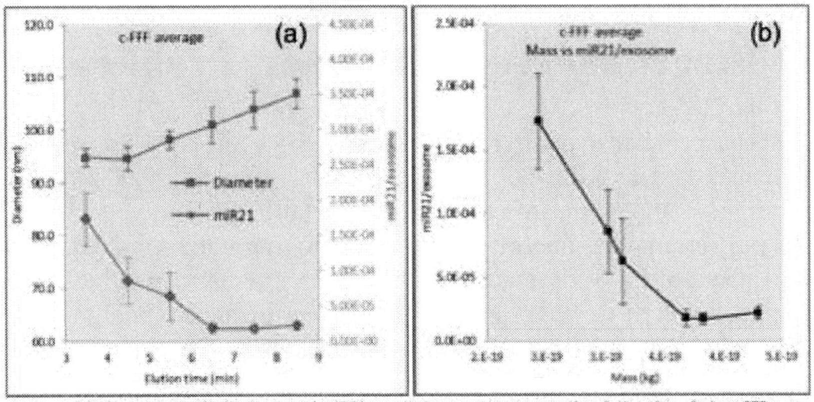

Figure 1: (a) Change in mode diameter and miR21 expression per exosome vs. the elution time during cFFF mass fractionation. (b) miR21 expression vs. average buoyant mass of exosomes in different mass fractions.

Overall, exosomes with smaller size and mass were found to be more abundant in miR21. A 10-fold difference was observed between the first and last fraction after cFFF and after AFFF. A 4-5-fold difference was observed when comparing the original sample used for field flow fractionation with the fraction being most enriched with exosomes containing miR21. The revealed molecular heterogeneity in EVs prior to mass-size fractionation explains the significant complexity in understanding the signaling role of EVs in health and disease.

568397 Cancer Progression As a Transport Problem

Approximately 600,000 cancer deaths occurred in the United States in 2018 but scientists have estimated the prophylactic and clinical measures developed since 1990 have avert 300,000 deaths a year. Researchers have made headway, but the complexities of cancer are very far being unraveled. Accurate prognostic models for disease progression play an important role in improving patient outcomes, yet aspects of tumor progression, such as the phenomena of dormancy and recurrence remain mysterious and our work appears to be the first quantitative model of it. We propose a population balance model that uses size-dependent parameters representing tumor growth, reduction and metastasis to predict how the distribution of tumor sizes in a cohort of patients changes with time. Size-dependence is a reasonable proxy for effects due to tumor genomics, because larger tumors tend to have acquired more adaptive mutations, and can acquire better access to nutrient supply. Such dependence means tumor growth rate is proportional to tumor surface area, volume or some combination.

The model transforms into an advection-diffusion transport equation in tumor size space. Having a diffusive term and variable parameters means tumors can have low Péclet and high Péclet number regimes of growth in the same patient. Transitioning from a low Péclet regime, where tumor size changes slowly and fluctuates randomly, to a high Péclet regime, where tumors grow or shrink quickly, is reminiscent of dormancy and recurrence in cancer progression. This model thereby provides mechanistic insight into one way recurrence can occur in patients. An important feature of using a distribution rather than some other measure for disease progression is the ability to capture rare events in the tails of the distribution. In disease progression, these rare events (i.e. an especially large cancer or a recurrence after many years of being apparently cancer-free) have the largest impact on the patient.

We have analyzed how the interactions amongst parameters affects tumor progression. For example, an effect resembling dormancy and recurrence predominantly occurs when growth is only slightly more size-dependent than cell death due to, e.g., immunity or chemo or radiation therapy. This means a wider range of sizes are in the low Péclet regime. Additionally, the side-dependence of the metastasis parameter can drastically alter the shape of the distribution curve at small tumor sizes.

To validate the use of the model for realistic data, we developed a fitting procedure to obtain rate parameters from experimental and clinical data. We successfully obtained parameter for literature data for hepatocellular carcinoma as well as our own far more extensive experimental data using a zebrafish melanoma model that captures the effect of immunity (and gender) on cancer growth. In the case of the zebrafish model, growth is more size-dependent than immunity, implying tumors in these fish can diffuse from a low Péclet regime to a growth regime. Using only a few size-dependent parameters, the model matches experimental data and recapitulates complex behaviors such as dormancy and recurrence that researchers find in clinical data.

577421 Reengineering the Tumor Microenvironment for Immunotherapy

Cancer cells promote the development of abnormal blood and lymphatic vessels through altered signaling and increased mechanical stress. This creates a hostile tumor microenvironment featuring hypoxia, acidity, and elevated interstitial fluid pressure. These abnormalities fuel disease progression, immunosuppression, and treatment resistance. Judicious use of antiangiogenesis agents that block the altered signaling can transiently 'normalize' tumor vessels. In parallel, because desmoplasia impairs vascular function by compressing vessels, normalizing the extracellular matrix also can improve vascular function. Here, I will present the progress made in understanding and applying the normalization concept to cancer and outline opportunities to improve the efficacy of emerging immunotherapies.

569770 Biogenic Metallic Nanoparticles. from Microbiological Biofactories to Nanometric Trojan Horses

Biogenic metallic nanoparticles. From microbiological biofactories to nanometric trojan horses.

Antimicrobial resistance to antibiotics (AMR) and cancer and two of the main concerns that the healthcare system should face nowadays. Current drugs and antibiotic treatments are becoming ineffective or have plenty of drawbacks related to misuse and overuse. Therefore, new alternatives are needed, and nanotechnology is rising as a powerful solution over time. How the nanomaterials are created has plenty of influence in their features and applications. Traditional synthesis of nanomaterials, taking knowledge from both physics and chemistry, is subjected to several disadvantages, such as the production of toxic-by-products and harsh conditions, as well as biocompatibility issues. Green nanotechnology is presented as a suitable answer, allowing the generation of nanostructures in a cost-effective and environmentally-friendly approach employing living organisms, such as bacteria, and biomolecules.

In this research, pathogenic bacteria (both antibiotic-resistant and standard strains of Gram-negative -such as Escherichia coli- and Gram-positive -such as Staphylococcus aureus-) and human cells (both cancerous -such as human melanoma and glioblastoma cells, and healthy cell lines, such as astrocytes of fibroblasts-) were used for the synthesis of different metallic -gold (Au), palladium (Pd), platinum (Pt), gold-palladium (AuPd) and gold-platinum (AuPt)- and metalloid -selenium (Se)- nanoparticles with sizes between 5 and 120 nm surrounding by an organic-derived coating. Bacteria and human cells were cultured in the presence of metallic salts under standard conditions, allowing the generation of nanoparticles through natural detoxification processes in a synthetic protocol that is followed using microscopy and spectrophotometric techniques.

After generation, the entities are purified and extensively characterized in terms of composition, morphology and surface chemistry, through techniques such as Transmission electron microscopy (TEM), Scanning electron microscopy (SEM), Fourier-transform infrared spectroscopy (FTIR), X-Ray Photoelectron spectroscopy (XPS), X-ray powder diffraction (XRD) and Raman spectroscopy, in order to elucidate the complex organic-derived coating surrounding the nanoparticles and the metallic/metalloid core. Nanoparticles were then used as antimicrobial and anticancer agents, with effects characterized through growth curve analysis, MTS experiments and colony counting growing assays, not showing significant cytotoxicity towards healthy human cells. To understand the mechanisms or action, Reactive Oxygen Species (ROS) analysis was accomplished as well.

The extensive characterization of the nanoparticles showed the presence of organic molecules -such as proteins and lipids- coming from the living organisms, with selenium nanoparticles with sized between 50 and 120 nm. Bacterial tests showed an unusual selective behavior in the antimicrobial effect of bacteria-mediated nanoparticles, indicating a dose-dependent inhibition when a bacterial strain X was treated with nanoparticles made by the same bacteria, while poor inhibition was found when different bacteria were used as a target. The nanoparticles showed a robust anticancer effect towards human melanoma and glioblastoma cells while remaining biocompatible towards the healthy cell lines.
On the other hand, noble mono- and bimetallic nanoparticles synthesized by human cells, with sized between 5 and 20 nm, were able to inhibit the growth of cell lines in a similar way that nanoparticles made by bacteria showed, with a certain degree of selectivity, remaining biocompatible. Besides, the production of nanomaterials induced a transition of the cells into a named zombie stage" or suspension toward the cells was not responding to chemical, physical or temporal degradation.

Interestingly, we observed that when the human cells were subjected to synthesis of nanoparticles, they were able to show a "petrification" process that allowed them to resist extreme environmental conditions, like starvation for several weeks, extreme high and low pH, heat, frozen and so on. Therefore, it was observed that the production of nanoparticles allow the cells to become extremely resistant to the environment. Besides, a study of how to stop the proliferation of cancerous cells was started, with the production of nanoparticles inducing inhibition of the proliferation of the cells in tissue models.

Therefore, we demonstrated that microbiological agents are successfully used as a synthetic machine for the generation of different metallic/metalloid nanoparticles of different compositions with biomedical properties. Therefore, they are presented as a suitable approach for the synthesis of nanomaterials in a green fashion, overcoming the limitations of traditional nanotechnology, and opening a new field for drug delivery and smart targeting of cancer and antibiotic-resistant infections.

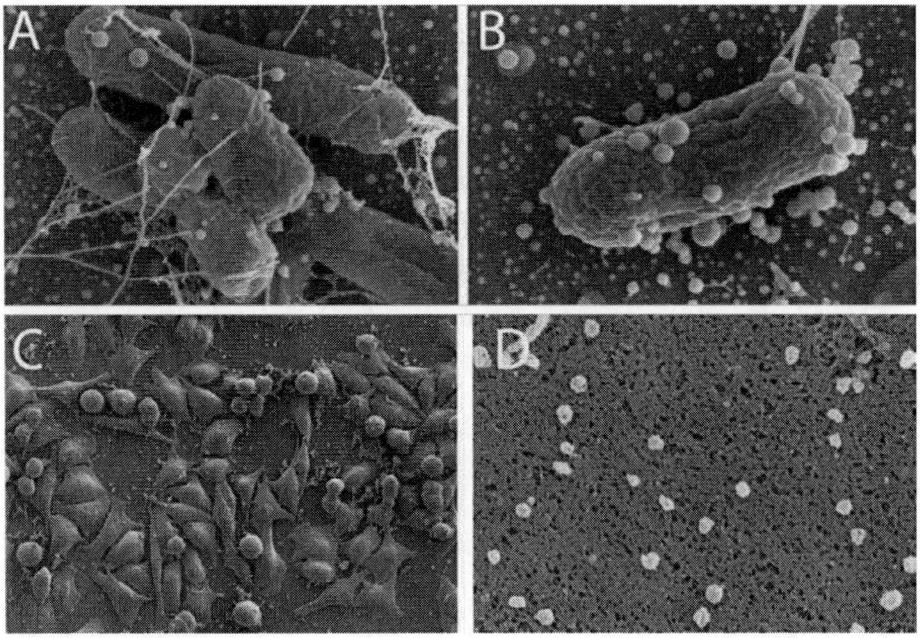

Figure 1. Selenium nanoparticles synthesized by MDR *Escherichia coli* (A, B) and cancerous cells (melanoma) syntehsized bimetallic gold/palladium nanoparticles (C, D).

577237 Characterization of Extracellular Vesicles By Atomic Force Microscopy

Exosomes and other extracellular vesicles (EVs) are molecular complexes consisting of a lipid membrane vesicle, its surface decoration by membrane proteins and other molecules, and diverse luminal content inherited from a parent cell, which includes RNAs, proteins, and DNAs. The characterization of the hydrodynamic sizes of EVs, which depends on the size of the vesicle and its coronal layer formed by surface decorations, has become routine. For exosomes, the smallest of EVs, the relative difference between the hydrodynamic and vesicles sizes is significant. The characterization of vesicles sizes by the cryogenic transmission electron microscopy (cryo-TEM) imaging, a gold standard technique, remains a challenge due to the cost of the instrument, the expertise required to perform the sample preparation, imaging and data analysis, and a small number of particles often observed in images. A widely available and accessible alternative is the atomic force microscopy (AFM), which can produce versatile data on three-dimensional geometry, size, and other biophysical properties of extracellular vesicles. In this presentation we guide the users in utilizing this analytical tool and outlines the workflow for the analysis of EVs by the AFM, which includes the sample preparation for imaging EVs in hydrated or desiccated form, the electrostatic immobilization of vesicles on a substrate, data acquisition, its analysis, and interpretation. The representative results (Figure 1) demonstrate that the fixation of EVs on the modified mica surface is predictable, customizable, and allows the user to obtain sizing results for a large number of vesicles. The vesicle sizing based on the AFM data was found to be consistent with the cryo-TEM imaging (Figure 2).

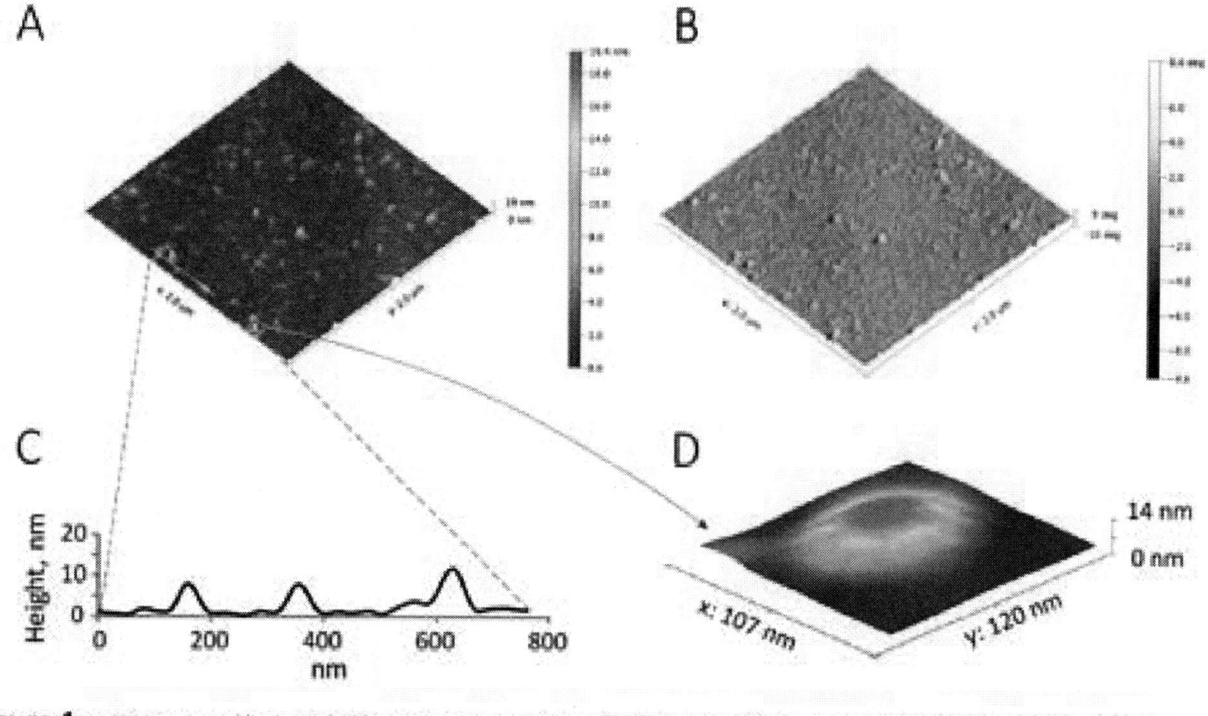

Figure 1: *AFM images of hydrated MCF-7 exosomes electrostatically immobilized on the modified mica surface. (A) The height image. (B) The corresponding AFM phase image confirms that the grains in the height image are soft nanoparticles, as should be expected for membrane vesicles. (C) The height data for the three vesicles crossed by the line shown in panel (A) illustrate a flattened shape caused by the electrostatic attraction of exosomes to the positively charged surface of the modified mica. (D) The shape distortion is apparent in an enlarged view the immobilized vesicle boxed in panel (A) and its cross section. The phase image of the same vesicle is shown in (E).*

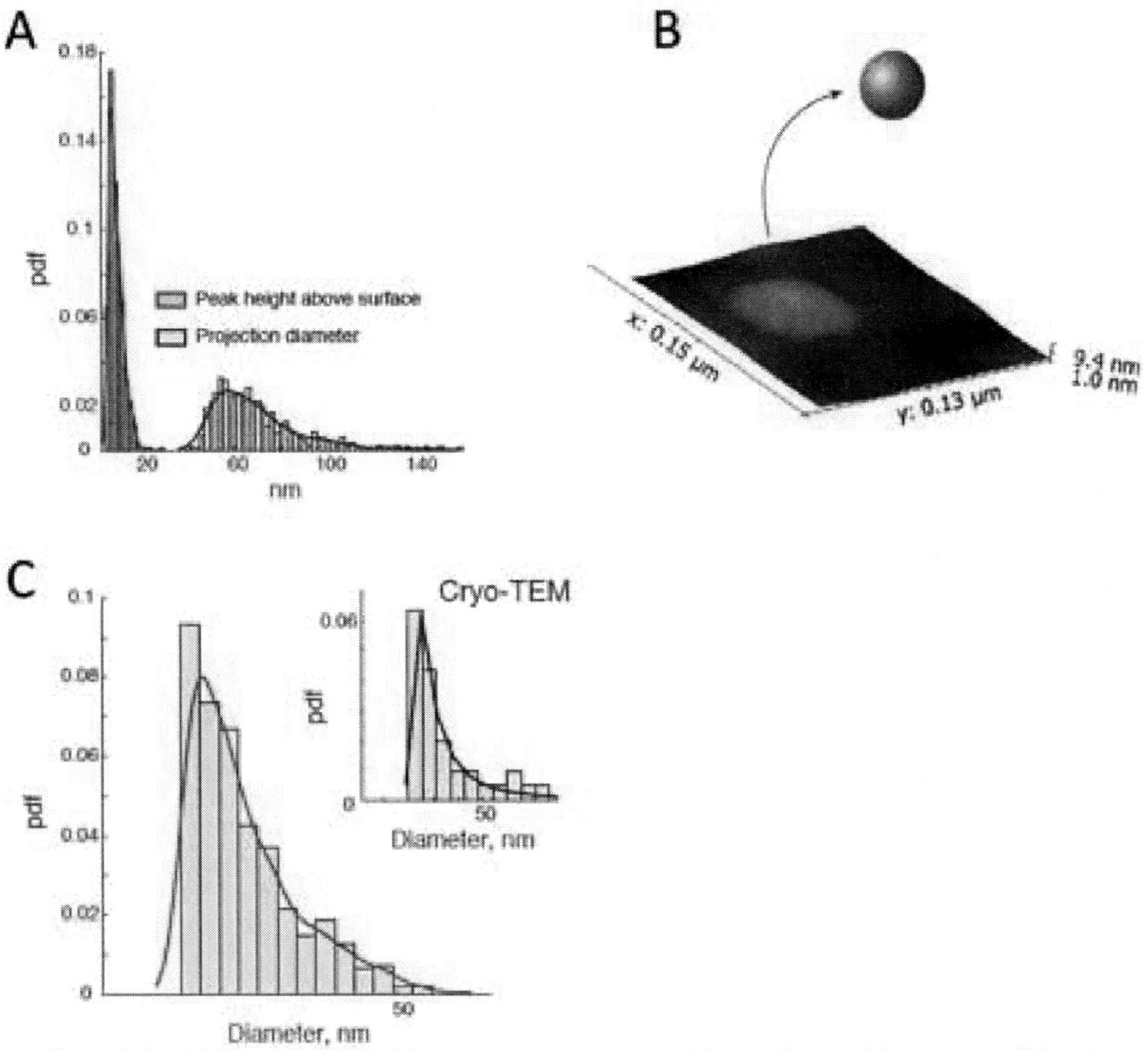

Figure 1: Dimensional characterization of hydrated vesicles immobilized on the surface and the estimation of their globular size in the solution. (A) The distribution of peak heights above the surface (red curve) has the mean equal to 7.9 nm. The area occupied by immobilized exosomes has 69.6 nm average diameter (blue curve). (B) AFM height image for one of the immobilized exosomes illustrates its highly oblate shape caused by electrostatic forces. The globular size of exosomal vesicles in the solution can be estimated by matching volumes enclosed by surface-immobilized and spherical membrane envelopes. (C) The size distribution of globular vesicles in the solution (red curve) was determined from the AFM data of 561 immobilized vesicles. The vesicle sizes in cryo-TEM images (blue curve) are consistent with the AFM results.

574970 Targeted Pegylated Mesoporous Silica Nanoparticles for the Delivery of Curcumin in a Pancreatic Cancer Animal Model: Inhibition of Both Tumor Growth and Metastasis

Background: Pancreatic cancer is one of the most difficult to detect and treat among all types of cancer. In fact, the five-year survival rate is less than 5% and is the only cancer, for which, the survival rate has not improved over the last 40 years. This is because nearly 80% of the pancreatic cancer cases are diagnosed when it spreads to nearby organs, a stage where surgery is not possible. So, there is a need to develop an effective way to deliver the anticancer drug to pancreatic cancer patients.

Method: Gemcitabine is the first-line chemotherapeutic drug for treatment of pancreatic cancer. Recent studies have shown that curcumin can potentiate the anticancer effect of gemcitabine in an animal model of human pancreatic cancer. The problem with curcumin is that, it is poorly water soluble, so one is not able to achieve its real potency. To counter this issue, mesoporous silica nanoparticles (MSN) were prepared by the sol-gel method, then loaded with curcumin (CUR) and coated with polyethylene glycol (PEG), with final conjugation via the targeting moiety transferrin (Tf) to synthesize MSN-NH_2-CUR-PEG-Tf, in order to target pancreatic cancer cells. Ability to load the drug in 2.5 nm pore diameter of MSN, bypasses the problem of poor solubility of curcumin, and in turn, potentially facilitate the controlled release of the drug, when it reaches the targeted cancer-site. In addition, the controllable and tunable pore diameter of MSN, over a range of approximately 2-6 nm pore diameter, helps in optimizing the relative balance between higher loading of drug and its controlled release rate from MSN and on to the target-site.

Results: TEM images reveal that, uniform spherical MSN were formed with an average particle diameter of 100 nm, which increased to 120 nm after coating of PEG over the external surface of MSN. This implies a 10 nm PEG coating thickness. BET analysis of gas-adsorption confirmed that, the drug was encapsulated in the pores of MSN, since surface area reduced from 669 to 80 m^2/g, and pore volume from 0.59 to 0.38 cm^3/g, for MSN and MSN-NH_2-CUR-PEG-Tf samples, respectively. From TGA we obtain the presence of 11.5% by wt. of curcumin in MSN.

Subsequently, *in vitro* cytotoxicity study on Mia Paca-2 pancreatic cancer cells proved that MSN-NH_2-CUR-PEG-Tf exhibited 2-folds higher cytotoxicity than free curcumin, proving that, nanoparticle mediated drug has higher cancer-cell killing capability. Confocal microscopy proved that the reason behind it is the 7 times higher curcumin uptake for MSN-NH_2-CUR-PEG-Tf, compared to free curcumin. The in vivo antitumor efficacy study in subcutaneous MIA Paca-2 mice model showed that MSN-NH_2-CUR-PEG and MSN-NH_2-CUR-PEG-Tf inhibit the tumor growth to a significant level and prevents its metastasis in different organs, like spleen, lung, liver, kidney etc. Furthermore, by using these curcumin nanoformulations, in combination with the standard chemotherapy regimen (gemcitabine), the cytotoxic efficacy increases many folds, in comparison to gemcitabine alone.

Conclusion: Hence, we propose that, targeted PEGylated mesoporous silica nanoparticles can be used as a carrier to deliver curcumin, in addition to gemcitabine, as an adjuvant therapy, in order to possibly improve the outcome of pancreatic cancer patients.

573961 Data Analytics and Optimization for Minimization of Chemotherapeutic Toxicity

Data Analytics and Optimization for Minimization of Chemotherapeutic Toxicity

Katherine Schmidt[1], Alex D'Aloia[2], Purnima M. Kodate[3] and Kirti M. Yenkie[2]*

[1]Department of Mathematics, Rowan University, Glassboro, NJ, USA

[2]Department of Chemical Engineering, Rowan University, Glassboro, NJ, USA

[3]Department of Pathology, Government Medical College, Nagpur, India

Cancer is a leading cause of death worldwide. Approximately 1 in 285 US children will be diagnosed with cancer before their twentieth birthday [1]. Leukemia, particularly ALL (Acute Lymphoblastic Leukemia) and AML (Acute Myeloid Leukemia), are the most common type of childhood cancers. Typical treatment consists of a long-term chemotherapy regimen, usually incorporating atleast one anthracycline, which has major limitations including the risk of cardiotoxicity [2], [3] and myelosuppression [4], [5]. From 2007 to 2013, the 5-year survival rate of children diagnosed with ALL was 88%.

Despite the high survival rate, most children will not live a normal life following the harsh treatment undergone by their bodies. Two-thirds of childhood cancer survivors will face at least one chronic health condition, while one-fourth will face a life-threatening toxic effect from the treatment afterwards in life [1]. Thus, it is of utmost importance to detect cancers at an early stage and implement suitable treatment strategies to avoid the disease progression as well as minimize any other health complications resulting from incorrect diagnosis and inefficient treatment.

One possible way of detecting the cancer early and identification of its correct sub-type is through the use of biomarkers. Some biomarkers are also capable of indicating additional toxic effects to the patient while undergoing cancer treatment. Especially, cardiac biomarkers are substances released in the blood when the heart is damaged or stressed [6]. These include proteins, enzymes, and hormones. Markers such as troponin proteins, natriuretic peptides, and myeloperoxidase have been studied as potential predictors of incidence and severity of heart failure in cancer patients [7]. In this work we have identified that myeloperoxidase is the main biomarker associated with leukemia subtype identification through the use of machine learning methods. Other variables studied include age, gender, periodic acid-Schiff, and total leukocyte count.

In addition to the early cancer diagnosis there is a need for appropriate treatment strategies that can simultaneously balance treatment efficacy and residual patient toxicity; herein lies an optimization problem. To this end, a systems engineering based mathematical model for cancer proliferation, tumor degradation, immunotoxic and cardiotoxic effects was developed and validated to facilitate the identification of clinical biomarkers responsible for long-term toxicity. Utilizing the information from the mathematical model, an objective function was formulated to minimize tumor size while maintaining the leukocyte counts and cardiovascular characteristics to the desired values. This resulted in an optimal control problem where the decision variable was the dosing values of the chemotherapeutic drug and its schedule for achieving the desired treatment objective. This problem was solved using two different methods: the maximum principle and discretized nonlinear programming [8], [9]. The dosing profiles predicted by both these method were then substituted in the model to check for treatment efficacy and the residual toxic effects. Our overall aim is to develop patient-specific treatment policies which can ensure a healthier life with no comorbidities for cancer survivors using principles of data analysis, mathematical modeling, cancer biology, optimization and control.

Keywords: Leukemia, chemotherapy, systems engineering, mathematical modeling, optimization

References:

[1] American Cancer Society, Surveillance Research, "American Cancer Society (ACS) Special section: Cancer in children and Adolescents," 2014.

[2] M. Florescu, M. Cinteza, and D. Vinereanu, "Chemotherapy-induced Cardiotoxicity," *Maedica (Buchar)*, vol. 8, no. 1, pp. 59–67, Mar. 2013.

[3] S. E. Lipshultz and M. J. Adams, "Cardiotoxicity After Childhood Cancer: Beginning With the End in Mind," *JCO*, vol. 28, no. 8, pp. 1276–1281, Mar. 2010.

[4] L. E. Friberg, A. Henningsson, H. Maas, L. Nguyen, and M. O. Karlsson, "Model of chemotherapy-induced myelosuppression with parameter consistency across drugs," *J. Clin. Oncol.*, vol. 20, no. 24, pp. 4713–4721, Dec. 2002.

[5] A. L. Quartino, L. E. Friberg, and M. O. Karlsson, "A simultaneous analysis of the time-course of leukocytes and neutrophils following docetaxel administration using a semi-mechanistic myelosuppression model," *Invest New Drugs*, vol. 30, no. 2, pp. 833–845, Apr. 2012.

[6] J. Mair, A. Jaffe, F. Apple, and B. Lindahl, "Cardiac Biomarkers," *Dis Markers*, vol. 2015, 2015.

[7] E. Christenson and R. H. Christenson, "The Role of Cardiac Biomarkers in the Diagnosis and Management of Patients Presenting with Suspected Acute Coronary Syndrome," *Ann Lab Med*, vol. 33, no. 5, pp. 309–318, Sep. 2013.

[8] U. Diwekar, *Introduction to Applied Optimization*. Springer Science & Business Media, 2013.

[9] K. M. Yenkie and U. M. Diwekar, "Comparison of different methods for predicting customized drug dosage in superovulation stage of in-vitro fertilization," *Computers & Chemical Engineering*, vol. 71, pp. 708–714, 2014.

AUTHOR INDEX

AUTHOR INDEX